REWIRED

REWIRED

Protecting Your Brain in the Digital Age

CARL D. MARCI, MD

Harvard University Press

CAMBRIDGE, MASSACHUSETTS

LONDON, ENGLAND

2022

Library of Congress Cataloging-in-Publication Data
Names: Marci, Carl D., 1969– author.
Title: Rewired : protecting your brain in the digital age / Carl D. Marci.
Description: Cambridge, Massachusetts : Harvard University Press, 2022. |
Includes bibliographical references and index.
Identifiers: LCCN 2021043329 | ISBN 9780674983663 (cloth)
Subjects: LCSH: Social media addiction. | Brain. | Human body
and technology. | Interpersonal communication.
Classification: LCC RC569.5.I54 .M37 2022 |
DDC 616.85/84—dc23/eng/20211012
LC record available at https://lccn.loc.gov/2021043329

For my children, Cam, Aria, and Luke

CONTENTS

INTRODUCTION

"The best time to plant a tree was twenty years ago.
The second best time is now."

—Proverb

I remember it like it was yesterday. As CEO and chief science officer of the company, I did not always attend our field studies. But some senior executives were scheduled to visit the lab to watch our research live, so I took the Amtrak into Midtown Manhattan that morning to oversee the experience. There was closed-circuit video set up in multiple small living rooms, so that we could view the first few participants. These young women were being paid $150 each to watch television for an hour while their smartphone behaviors and eye movements were recorded for analysis by our team back in Boston. Half the women were asked to leave their phones in another room, so that we could compare their behaviors with those of participants who were able to keep their phones.

Immediately I noticed something strange about the first participant in the group. She was visibly distressed without her smartphone—frowning, fidgeting, and squirming in her seat. She clutched the couch like she was about to jump out of her skin. At first I thought she was just getting settled into the unfamiliar surroundings, or maybe she was having an anxiety attack. But the odd behavior continued. I then wondered whether she was using drugs, which sometimes happened in our studies. But the staff did not notice any strange behaviors in the moments prior to the study.

Her odd behavior was getting worse as time passed. I started to think she was going to leave early. After about ten minutes, that is exactly what happened. She jumped up, ran out of the room, grabbed her smartphone and other belongings, and without payment or comment she left the facility. She was not the only one.

The fall of 2011 was early in the smartphone revolution. The research team at Turner Broadcasting was paying us for a custom neuroscience-informed study that would shed light on a new phenomenon: the simultaneous use of a second screen (smartphone or tablet computer) while watching live television, which media and advertising executives often call the *first screen*. We expected viewers would use their smartphone when it was available, significantly distracting them from watching the advertising on television, but viewers without smartphones would be more attentive to the TV ads. In fact, that's just what we found. But we did not expect some of the responses we saw from several participants deprived of their smartphones.

We had enlisted women between the ages of eighteen and twenty-two. They were all heavy media multitaskers: they told our recruiters that they frequently watched conventional TV and used their smartphone at the same time. Each participant was randomly assigned to one of two groups. Women in the first group were asked to watch an hour of television with their phones available. They received simple instructions: "use your smartphone like you would at home watching television." Those in the second group were asked to leave their smartphones and any other technology in a secure waiting area for the duration of the study.

Of the thousands of research participants we have studied over the years, none surprised me like this. Healthy, young people were in physical and emotional distress, apparently owing to smartphone deprivation. It was not just one person. The same odd behavior I witnessed in the first participant occurred in several others. Of the group that was allowed to keep their phones, no one left without finishing the one-hour study. They all calmly watched TV while periodically checking their phones—and then got paid.

It made me wonder what might be happening. Could it be that these young women were going through a form of withdrawal caused by separation from their smartphones? It occurred to me that they were behaving like the rats placed in Skinner boxes for drug studies in the 1950s, which I had watched on films as a premed and psychology undergraduate. B. F. Skinner was the father of behavioral theory. Part of his fame came from putting lab rats in a small box where they could press a lever to access sugar water or a drug such as cocaine or heroin. The lab rats quickly learned to choose the drug over the water and, in time, became addicted. Then Skinner would take access to the drug away from the rats and watch their behavior change. The deprived rats would fidget and squirm in their box, suggesting a level of agitation. This was an early insight into the behavioral signs of withdrawal that result when the addicted brain is deprived of drugs of abuse.

Were these emerging smartphone habits also addicting? When I watched that young woman squirm, I came face-to-face with a new reality, one far scarier than our big media clients were interested in. Turner Broadcasting was very concerned about the ways in which this powerful mobile media technology was changing television consumption habits and about the rapidity with which those changes were happening. But these concerns were primarily related to advertising revenue. I saw other problems. I had the privilege of being at the leading edge of media research. I saw people's habits changing in front of my eyes. And as a trained physician and psychiatrist who had studied the brain for years, it seemed to me that advertising revenue was not the most serious of our worries.

Our world is changing. Fast. The way we work. The way we travel. The way we entertain ourselves. And, importantly, the way we interact and communicate. In fact, the pace of change is so incredibly fast that it can be hard to understand what, and just how much, change has happened over the last decade. Our brains aren't wired to understand this level of complexity. We store and recall what's relevant to us here and now. Our ability to place things in a historical context and understand them is limited when change occurs at high speed.

Our brains are imperfect information processors, yet we have become accustomed to thinking of our brains as computers. So powerful is this metaphor that we no longer cringe at the notion of being *wired* or of *rewiring* ourselves. But as computer processing speed, availability, and capacity increase, it appears that human brain processing speed, availability, and capacity may be decreasing. This has important consequences. This book is about those consequences.

The modern media landscape raises important questions for children, adults, parents, caregivers, educators, politicians, and researchers. Understanding when to introduce what type of media experience to young children and how to set limits is of growing concern for all of us. Interactive programs, whether educational apps or casual games, are increasingly popular. Some are too complex or too new for academic researchers to assess beyond preliminary insights. This forces us to look back at prior research with old media as a guide to interpreting new research on the impact of new media. And with the lens of modern neuroscience, several concerning themes become clear.

Our brains are constantly changing. Neural rewiring occurs from the day we are born to the day we die, albeit in different ways over the course of our lifespan. All experiences shape who we are and who we are becoming. Small experiences have a small impact on the brain. Large experiences have a large impact on the brain. Some experiences shape our brains for better and some shape us for worse. Our experiences with mobile media, communications, and information technologies are no different. Screens are now ubiquitous in the classroom, in the workplace, and in every facet of our lives. The mobility and accessibility of screens foster small and large experiences every day, many times a day. These experiences rewire our brains as they create new habits, influence our mental and physical health, define who we are (or who we pretend to be), and affect how we relate to one another.

The "rewiring" metaphor that is used throughout this book is not meant to imply that rewiring our brain is necessarily good or bad. It is meant to remind us that when we change our behaviors, we change our brains. Period. When we are blind to massive changes and more reactive than proactive when faced with the power of the supercomputer in our pocket, this rewiring can result in negative consequences. When the consequences related to global shifts in the use of these devices are ignored or minimized, they will contribute to unhealthy outcomes at all levels of a society increasingly dependent on the very same devices. These are some of the themes we will explore in this book.

Part 1, Wired: Connected Brains, sets the stage. I review some milestones in the history of media and advertising, which created the conditions for the rapid ascent of what we euphemistically call the smartphone. I then introduce some key concepts about the prefrontal cortex, the most sophisticated and intricate region of the human brain. The power of the prefrontal cortex is clear in both the evolutionary history of the species and across the lifespan of the organism, as individual humans develop from childhood through adulthood. The prefrontal cortex is the key to our humanity. It gives rise to executive function and plays a critical role in our relationships—relationships that emerge because we are fundamentally wired to connect as social creatures. The prefrontal cortex is also the region of the brain most vulnerable to the impact of our changing media and technology behaviors.

At the same time, the prefrontal cortex is our best defense against the many new sources of threat in the digital age. This powerful bit of brain protects us from succumbing to superstimuli and other tricks and attractions online, with their endless rewards and compulsion loops that are literally an arm's reach away. A healthy prefrontal cortex helps us interpret and manage the responses we receive from our more primitive reward and emotion centers. It is the key to forming relationships and to a healthy self-identity. The prefrontal cortex helps us achieve success at school and at work.

Collectively these capabilities make the prefrontal cortex an essential arbiter of impulse control. The prefrontal cortex guards us against poor decisions and thereby mitigates the impact of unhealthy behaviors that can lead to unhealthy habits and addiction. As we will see, if the prefrontal cortex becomes even slightly compromised due to stress, neglect, fatigue, media multitasking, information overload, or succumbing to the superstimuli and disinformation that are ubiquitous online, we put ourselves at great risk. Our risk of forming unhealthy habits and ultimately addictions increases just as our ability to make healthy judgments and decisions for ourselves, our families, and our society decreases.

This is the theme of part 2, Rewired: Assaulted Brains. This section takes us through the many consequences of the rapid ascent of smartphones in our lives. I examine the impact of this changing world from the beginning of the lifespan, identifying major impacts of smartphone use on the prefrontal cortex from infancy onward and exploring the repercussions on the growing brain and person. The goal is to create a framework for understanding new recommendations for parents and all people, based on the best available science.

Each life stage presents the brain with unique challenges. One of these challenges, the one presented to the young study participants at the Time Warner Media Lab in New York, is media multitasking. Almost all of us have embraced this new habit, but it has its dangers. I will describe research showing how multitasking tricks people of all ages into thinking we are being more productive. In fact, multitasking reduces speed and efficiency, so we accomplish less while potentially working harder. Turning exclusively to adults, I then explore the fine line between habits and addictions in the brain and discuss the consequences of ignoring our new habits, which can have a profound impact on our relationships, our mental health, and our physical well-being.

But while our digital habits create hazards at every stage of life, we are not without opportunities to protect our brains. In part 3, Beyond Wired: Better Brains, I guide readers toward a comprehensive digital literacy, explaining how to identify digital-age problems in ourselves, our coworkers, our friends, and our loved ones. Included are ten empirically driven recommendations to help protect our brains in general and the prefrontal cortex in particular. Against the onslaught of the moment, we should keep in mind the meaningful solutions we can bring to bear.

It won't do to just discard our digital lives. We can't. And there is a lot to like in these technologies. Smartphones and related tools offer portable powerhouses of information, communication, media, and commerce. They enable and encourage a constant connection to entertainment, news, work, education, friends, and family. They literally put the world at our fingertips while rarely leaving our side.

The benefits are obvious and indisputable, and so are not really my focus. Instead, I want to take a sober look at the costs and negative consequences as well as the opportunities we have to mitigate the downsides. The rapid adoption and unprecedented behavioral changes these technologies inspire explains the growing concern about the impact of these technologies—even as we continue to adapt to their ubiquitous presence. New habits, new forms of content, and new ways of engaging with mobile media, communications, and information technology are changing our children's developing and our adult brains in meaningful ways. As we explore the impact of smartphones and other technologies on the developing and ultimately the adult prefrontal cortex, my goal is not to cover every potential topic in a comprehensive manner. Rather, I aim to highlight research that speaks clearly to the benefits of digital literacy—for everyone.

Because of their extraordinary benefits, and despite the many challenges they create, smartphones are here to stay. We all need to work together to put up some stoplights and warning signs on what we once fondly called the information superhighway. And we need to learn how to use these rapidly evolving technologies in service of healthier versions of ourselves, while avoiding their detrimental consequences. As a society we need to be more proactive than reactive in our approach to a changing world of mobile technology with its constant connections, endless rewards, and very real consequences.

COVID Considerations

The COVID-19 pandemic altered nearly every aspect of our lives, even halting some activities critical to well-being. But beyond the staggering death tolls, economic strife, and disruptions in how we work, learn, and play, the impact on the mental health of children and adults created its own set of problems. Think about it: more time indoors, social isolation, remote work and school; the cessation of live entertainment, travel, and social gatherings from grade school to graduations. Many of us have spent a lot more time on screens since March 2020, a phenomenon with worldwide implications for mental and physical health.[1]

We have even developed names for new problems like "Zoom fatigue" and "pandemic panic." We learned to sit all day on video conferences, constantly messaging exhausted and burned-out colleagues on Teams, Slack, and WhatsApp. We've taken bathroom breaks hoping we hit the right button to mute the sound and turn off the camera. And after peeling our tired and strained eyes off the screen at the end of the day and taking a short break for dinner, it's on to the Chromebook or the Seesaw app to help the kids with homework. After finally crawling into bed, we wonder if we ever got this tired before the pandemic—it is hard to remember. That is, if we have been lucky enough to have a job that allows us to work at home and the resources to educate our children remotely.

Before the pandemic, we spent a lot of time online consuming media, communicating with each other, and playing games. New habits were formed, some of them bordering on addiction, all of them with consequences. Since the pandemic, things have only gotten worse. Shifts in complex human behaviors are always hard to understand, and researchers are just beginning to sift through the data. But if there were rising concerns before COVID-19, alarm bells should be ringing in the aftermath.

The impact of the virus on our collective screen behaviors is easy to see but hard to comprehend. When comparing time spent online accessing news and current events by US adults in March 2019, before the pandemic, and March 2020 at its onset, we see a nearly threefold increase. Time spent streaming video hit an all-time high on Easter weekend 2020, as COVID-19 case counts increased and Americans looking for a distraction from the news sought out entertainment online. One global estimate finds that children spent twice as much time staring at screens in May 2020 as in May 2019. This COVID effect, as some have called it, sees more time spent on gaming, social media, and apps associated with schooling. The implications of a year and more of altered screen behavior around the world are potentially staggering.[2]

How staggering? As is a constant refrain in this book, correlation is not the same as causation. However, when we consider the shifting use of media, communications, and information technology during the pandemic, and as we see how the lives of so many adults and children have been interrupted, it is not surprising to hear statistics about a growing mental health crisis, whose consequences may linger longer than those of COVID-19.

Research shows unprecedented levels of stress during the pandemic, and there have been calls for more mental health resources from almost every corner of society. UK government surveys show that 10 percent of British adults reported symptoms of depression between July 2019 and March 2020—compared to 19 percent in June 2020. And in the United States, the proportion of adults reporting symptoms of either anxiety or depression rose from 11 percent in June–December 2019 to 42 percent in December 2020.[3]

This book was written before the pandemic drove hundreds of millions of people around the world inside their homes in the name of social distancing, leading to still-greater dependence on media, education, and productivity technologies. Even as the crisis resolves, its impact on screen time, media multitasking, and the behavioral changes associated with the digital age are likely here to stay. After all, when it comes to the influence of digital technologies in our lives, COVID reinforced trends long underway and has only made the problems investigated here more urgent.

PART I

WIRED

Connected Brains

1

MEDIA MATTERS

Scanning the news online, a headline catches my attention: "Man, Distracted by Electronic Device, Identified after Falling to Death at Sunset Cliffs." It is shocking and tragic.

The authorities identified the victim as a thirty-three-year-old Indiana man who was visiting friends in San Diego. Witnesses said he was not watching where he was going and was obviously "looking down at the device in his hands." Distracted by a smartphone or similar handheld digital device, he lost his footing and fell sixty feet off the cliff edge. He was pronounced dead at the scene. Authorities used the opportunity to remind citizens that we need to be mindful of our environment when using digital media devices, whether we're tempted to text while driving or enjoying a hike.

This incident from 2015 is not the only case of digital distraction leading to death and other physical mishaps. That same year, two tourists slipped down the steps of the Taj Mahal while posing for a selfie; one died of head injuries, the other fractured his leg. Another man was so focused on taking a photo of himself during the annual running of the bulls in Spain that he failed to get out of the way of the animals and was gored to death. A woman fell off a pier into Lake Michigan while typing into her phone. Another woman fell into a water fountain inside a mall while walking and using her smartphone. Neither was seriously injured, but the fountain diver was mortified when the mall security team shared video of her fall on social media. You can hear the security guards laughing at her as they watch the video repeatedly at multiple angles. She is seen in a subsequent interview in tears as the video goes viral, amplifying her grief.[1]

Some of these, to be sure, are extreme examples. But they illustrate how serious the consequences of tech behavior can be. Many of our new media, communications, and information technology habits are resulting not only in more accidents from distractions but also reduced productivity, disrupted relationships, increasing mental and physical health problems, and public safety issues that we need to take seriously and address as a society. In fact, many of the consequences of our new behaviors are so severe that experts are starting to describe them with the language of addiction. This suggests that the rewards from new applications of technology are so powerful that they are rewiring our brains to automatically respond to every app ping and message ring, no matter the risks that may be associated with that behavior in a particular context.

Despite the many advantages of these technologies, there are signs that lots of people recognize the downsides. A study of consumers across nine countries by the Ford Motor Company found that 78 percent of women think technology is contributing to sleep deprivation and 63 percent of adults agree that technology is making them more impatient and less polite. The study also found that some 80 percent of adults think social media is more "optics" than substance—that is, people perceive online representations as commonly false or self-serving. The report calls this a "tech spiral" caused by continuous connectivity, changes in the way we communicate, and countless hours consuming a nonstop stream of entertainment, social media, advertising, information, and too often disinformation.[2]

All. The. Time.

The Ford report concludes that in response to our changing technology world, there is now clear and undeniable evidence that we have a love-hate relationship with our mobile media, communications, and information devices. "Technology continues to unfold at breakneck pace, but it's also prompting greater reflection on the impact on our lives," states Sheryl Connelly, Ford's corporate futurist, in the report's introduction. She cites the mounting number of people claiming that our increasing reliance on mobile gadgets has negatively affected society.

We are getting more distracted, divided, and depressed. We are changing the nature of our social bonds and our brains. We are all less engaged in intimate ways with the people and experiences that we used to care about. Change comes at a price. Our collective tech-life balance needs adjustment as we wrestle as a society to define a new form of digital literacy. As we adopt modern mobile technologies and embrace their myriad benefits, we need to be cognizant of the growing consequences of how we adapt to their use.

How can we begin to understand the drivers of these massive shifts in our behaviors as the modern smartphone rewires our brains and alters our lives? Before we look forward in an attempt to restore balance, we need to look backward and understand how we got here.

Our Changing World

In order to appreciate the massive shifts in our relationship with modern media technology today, we need to answer an important question: How did media come to dominate our lives?

One compelling theory, offered by sociologist Todd Gitlin, suggests that in the early twentieth century, as the gains of the Industrial Revolution were being realized, American workers and employers made an implicit bargain. On the one hand, workers could take advantage of increased productivity by working less and gaining leisure time. Many European countries went this route, choosing longer vacations over increased paychecks. On the other hand, workers could choose higher wages. Gitlin contends that American business leaders and laborers chose to pad their pocketbooks.

With more money and new technologies enabling faster production, corporations realized they had an opportunity to boost profits. As it became relatively easy to generate new products while allowing older ones to descend into obsolescence, companies could repeatedly introduce improved goods and services. The challenge was to ensure that there was demand for them. To generate mass consumption, mass advertising was needed. Corporations rolled out a constant stream of messages that began to seep into and shape our collective imagination. As Gitlin notes, "Pictures of purchasable utopias proved pervasive. Even while protesting how many hours they worked, workers came to prefer the pursuit of happiness via commodities to the pursuit of leisure."[3]

What made these pictures of supposedly purchasable utopias ubiquitous was a new kind of media platform that provided the necessary channels for advertisements. In the early days, that platform comprised mainly ad-supported radio programs, newspapers, and magazines. By mid-century, television came along and reshaped our living rooms with easily accessible moving pictures of live news, sports, and entertainment. Early television programming with paid sponsorships on just a few broadcast networks generated enormous audiences and enormous profits for media companies and the firms that advertised with them. Over time, we radically changed our behaviors, consuming far more media than ever before and doing so in new ways and new contexts.

As television evolved, mass marketing took another great leap. The advent of cable as an alternative to and eventually replacement for terrestrial broadcasting brought about the proliferation of entertainment networks and an explosion of programming content for all ages. More content meant more media consumption, larger audiences, and more advertising revenue. But gradually these audiences fragmented, giving rise to increasingly sophisticated media and market research and the notion of "segmenting" an audience not only by age, gender, and geography but also by taste, preference, and income.

But even as media technology changed, there were a few constants between the 1950s and the 1990s. The telephone was a communication device, the computer was a productivity device, and the television was an entertainment device—all fueled by growing consumer demand and a growing ability for advertising to fund and feed that demand.

The rise of the personal computer, the internet, and then the smartphone undermined these constants. Now all screens increasingly are all things to all people. As a former media and market research executive, I have seen firsthand the transformation of the modern media consumer and the impact of smartphone technology on the ways we engage with media platforms and content. In the not-so-distant past, mass media was a shared, common experience. Organizations sought to reach a large and diverse audience with mass marketing. Today these same organizations invest in targeted marketing. The goal is to influence the individual consumer's choices through a personalized appeal, not to blanket the airwaves with a single message in the hope that lots of consumers will notice.

We can now choose what we watch, where we watch, when we watch, and how we watch in ways unimaginable even a few years ago. And the result of all these choices is a huge increase in time devoted to media. Consider statistics from the Nielsen Company, the world's oldest and largest provider of media viewing and purchasing data, where I served as global chief neuroscientist and executive vice president for several years during a time of tremendous change. Nielsen measures media usage across ten delivery devices: live TV, streaming programming, AM and FM radio, smartphones (apps and web), personal computers, digital video recorders (DVRs), DVD players, tablets (apps and web), video game consoles, and what are counted as "other" multimedia tools. Results are presented quarterly in Nielsen's *Total Audience Report*, where we see that in less than two decades, there has been a massive shift in media consumption time and the types of devices used. According to Nielsen, in 2002 US adults consumed an average of about forty-eight hours of media per week, mostly via TV, radio, and videocassettes. By 2015 usage had increased to nearly sixty-four hours per week. By 2016 US adults devoted a staggering 74.5 hours per week—10 hours and 39 minutes per day—to media consumption across nine different media devices. In 2018 the figure topped eleven hours per day. Americans now spend the equivalent of nearly two full-time jobs consuming media each week![4]

How did Americans manage to find an extra thirty-plus hours a week over the last two decades to consume media? The answer: mobile technologies and the rise of media multitasking. While television continues to dominate viewing behaviors of children and adults, 2011 saw a decline in the amount of live television viewed on average per person per week in the United States. This was the first such decline in the history of Nielsen ratings, which had been measuring viewership since 1950. Even so, overall media consumption increased, with the largest gain coming from the use of portable personal media devices that access content through the internet. Such devices more than made up for the drop in TV viewing, and chief among those personal media devices was the smartphone.[5]

The New First Screen

Today we tend to forget the first relatively "smart" phones with email and internet capabilities, notably the BlackBerry by Research in Motion, which started as a two-way pager in 2002, and the Palm Treo, the first mobile device to send emails and allow users to dial from a list of contacts. But many of us remember the atom-splitting moment that came in 2007, when Apple introduced the pocket-sized iPhone with a sophisticated and intuitive touchscreen, an advanced operating system, and the most impressive software and hardware functionality the world had ever seen. Others followed, including Google, with the Android operating system that would compete with Apple's iOS, and Samsung with a suite of phones going toe-to-toe with Apple's hardware.

This new generation of smartphones possessed capacities that attracted users everywhere. Most important among these capacities was access to the internet. Yes, the machines were fashionable and easy to use, integrated valuable technologies like text messaging, and played music on the go. But it was the continuous connection to the internet and access to the functionality of third-party software applications—that is, apps—that drove the rapid adoption and use of smartphones. The always-on link to the web granted access to this growing universe of software and enabled users to consume any media content that was also online. With these tools, users could also be producers of media content. As new media producers, they too could sell content and valuable "real estate" to advertisers. But mostly, with smartphones in their pockets, users were easy targets for advertising. As Jack Wakshlag, former chief research officer at Turner Broadcasting, puts it, "Smartphone mobility creates new markets of time that extend the places and ability of people to watch video and engage with advertising."[6] All this further drove the rising popularity of smartphones.

Sales of the iPhone exploded, and in August 2011 Apple became the most valuable company in the world, topping Exxon Mobil. It was a symbolic transfer of power from oil and energy to information technology, from heavy industry to commerce in communications, media, and data. In 2018 Apple broke another record and became the first company to reach a market capitalization of $1 trillion. And between these events, in the summer of 2016, the smartphone quietly reached another important milestone, penetrating over 80 percent of the mobile phone market—just one year prior to the iPhone's tenth anniversary.[7]

Indeed, smartphones have set a new standard as the technology with the *fastest* rate of adoption in the recorded history of humans. Historians of technology look at changes in adoption rates, or penetration, as good indicators of the pace of uptake, allowing them to compare growth rates of vastly different technologies. When thinking about transformative technologies, the move from 40 percent penetration to 75 percent penetration is considered an important benchmark. It took electricity and the telephone more than fifteen years to go from 40 percent to 75 percent penetration in the United States. Personal computers and the internet took about ten years.[8] Television was for a long time the champion, clocking in at five years. Data from Nielsen and others show that smartphones made the leap from 40 percent to 75 percent growth in just three years. An astonishing achievement.

The mass penetration of the smartphone changed the media market in ways few television executives or major brands could have anticipated. Thus, while growth rates for smartphone sales have slowed as the market becomes saturated, what has not slowed is spending on digital advertising, powered by the endless consumer appetite for new content and apps and by the simple fact that mobile devices are the best tools with which to engage users. So it was that in 2017 online ad spending surpassed television ad spending for the first time.[9] In the advertising world, mobile devices connected to the internet are the new first screen, replacing television's decades of dominance.

While I believe that connectivity is the key to the success of smartphones, I don't want to discount another monumental feature. Sophisticated smartphone cameras have changed our relationship to photos, videos, and our image of ourselves and others—forever. I watched with marked interest how my two-year-old daughter saw herself on my iPhone, doing something she did a moment ago that was serendipitously captured on video. She halts, smiles inquisitively, and then laughs shouting, "Daddy, again, again!" Wishing to feed her curiosity, I oblige.

What is the impact of seeing oneself at such an early age on video or in pictures that are instantly available to ourselves and others? Young people today, with Snapchat filters, PopSockets, and broad social media access, no longer view video in the same way as teenagers of even a few years ago. What does this do to their emerging sense of identity? As Steve Hasker, a former Nielsen executive, puts it, "Teenagers today are more interested in watching themselves kick a goal than watching [soccer star] David Beckham kick a goal."[10]

It's not just children and teenagers who are changing their behaviors and habits—it's all of us. The average US household now has twenty-five connected devices, including smartphones, laptops, desktops, streaming devices, smart TVs, gaming systems, fitness trackers, and more. The result is more mobile media consumption, digital communications, and information processing than ever before.[11] The same is true the world over. India, China, and the countries of Africa are becoming "mobile first" and "mobile only" cultures, as the smartphone becomes the screen of choice over television and personal computers.

We are also using our smartphones to access mobile media content in a wide variety of settings that were unimaginable only a few years ago. This gives rise to the other new and consequential behavior I have mentioned, which will be a major theme throughout the book: media and technology multitasking. We can now use media, communications, and information technologies in the store, in the park, in bed, walking down the street, or while sitting in our cars—too often while driving.

What was unimaginable a few years ago today is indispensable. Again, this is not just a US phenomenon. While the United States is a leader when it comes to overall media consumption, the internet and smartphones are propelling similar trends around the world. As overall global use of traditional media like TV, radio, newspapers, and magazines declines, global consumption of media and entertainment increases dramatically through the use of mobile media devices.[12]

What drives us to consume so much media? As early media researchers put it in 1973, we use media for many reasons: "To match one's wits against others, to get information and advice for daily living, to provide a framework for one's day, to prepare oneself culturally for the demands of upward mobility, or to be reassured about the dignity and usefulness of one's role."[13] In other words, media fulfills multiple perceived personal and social needs.

Today these needs have only multiplied. Now we use media not only to be informed, entertained, and validated, but also for self-expression, companionship, and emotional stimulation. The uses, gratifications, and rewards that we get from media are more diverse and more emotionally rich than ever. The ability of mobile media, communications, and information technologies to fulfill so many needs and offer so many rewards in so many locations drives new habits around mobile technology that are rapidly changing how we live, love, and labor. Despite the undeniable benefits, these new habits are altering our brains and behaviors in measurable ways and with unanticipated costs and consequences.

Media as a Mood Regulator

In the fall of 2011 my company, Innerscope Research, took on an exciting new project. We were pioneers in the use of neuroscience-inspired technologies to help answer questions about media consumption and marketing in a changing world. We were asked to develop a revelatory study for Time, Inc., publishers of *Time* and *People* magazines. Betsy Frank, a Time executive and leader in media research, wanted to understand the rapidly evolving media landscape and get a handle on how different generations of consumers were using new digital devices. To really understand these changes, she needed to employ new research methodologies.

Innerscope designed a study that would passively monitor media consumption in the lives of two groups: digital natives and digital immigrants. Digital natives were defined as younger adults, born after 1990, who grew up in the digital age and never knew a world without the internet. They were perceived as being more interested in digital media platforms connected to the internet and less interested in "traditional media"—that is, media without a connection to the internet. Digital immigrants were born before 1990—perhaps, like me, well before. They recalled the days of traditional media's dominance, before the internet existed or was widely available. Digital immigrants had to adjust to the digital age.

The research participants were all in the Boston area. They all owned a smartphone and an iPad and were paid to live their lives as usual except that they would be wearing special point-of-view glasses with cameras that recorded every moment of the one-and-a-half-day testing period, and they would wear a biometric belt to measure their physiological responses to their environment. It was a challenging and technical study. After we completed the fieldwork, we moved on to the daunting task of coding and analyzing over 300 hours of unique data from this first-of-its-kind study. We called it, "A Biometric Day in the Life."[14]

The results changed the way I think about the relationship between emerging media habits and their impact on our behaviors and our brains. In terms of time spent with media, consistent with national statistics, we found that both digital natives and digital immigrants consumed a lot. On average, participants spent two-thirds of their nonworking hours consuming media of some sort. And while both groups consumed about the same number of hours of media throughout the day, true to their moniker, digital natives spent a disproportionate amount of their time on digital devices, while digital immigrants spent more time with newspapers, magazines, radio, and television.

This was not surprising. What was surprising was the way participants paid attention to media. During the analyses, we noticed something peculiar in their media behaviors: participants in both groups switched their attention among media platforms with surprising frequency. While rapid switching between media devices was common to both groups, I wondered if it was more prevalent among digital natives. In other words, was there a difference in *media attention span,* a term we coined, between the two groups?

We realized we were witnessing the evolution of media multitasking from a unique perspective, and what we found generated headlines.[15] The older digital immigrants switched media devices an average of seventeen times per hour, translating to an average media attention span of about three and a half minutes. In comparison, the younger digital natives switched media devices on average twenty-seven times per hour, averaging nearly every two and one-quarter minutes. This is an astonishing 60 percent increase in the rate of switching, making for a huge drop in the media attention span across a single generation.

The biometric portion of the study was equally fascinating. We used biometric devices that capture real-time, moment-by-moment changes in the participant's heart rate and skin conductivity—two physiologic indicators of emotional intensity. We also time-locked the biometric data to the video from the cameras in their glasses. This allowed us to analyze how users' emotional responses fluctuated as they consumed different types of media throughout the day.

As expected, we found that, on average, digital natives were more emotionally engaged with digital media platforms and digital immigrants were more emotionally engaged with traditional media platforms. This is not surprising given that digital natives spent more time using digital platforms: our emotions drive our behaviors.

What was surprising was that both digital natives and digital immigrants experienced a greater range of emotional responses when using a single platform for sustained periods than when media multitasking. This suggests that when we are focused on a single platform, our emotional response is more dynamic: the highs are higher and the lows are lower. In contrast, media multitasking constrains both the highs and the lows, but the average emotional response is more intense. The result is a flatter but more consistently elevated emotional response.

This got us thinking. What drives consumers to switch media platforms so frequently when doing so results in blunted emotional peaks? What is it about media multitasking that the brain finds rewarding? The answer was suggested in the data but emerged more clearly after several other studies in the more controlled environment of our lab. We found that people switch from one media experience to another when their emotional response begins to drop in intensity, suggesting lower levels of arousal. In other words, we found that media consumers switch media platforms when they are bored.

What we have come to learn is that our mobile devices provide so many opportunities for rewards, so many motivations to engage, that they are inexhaustible vehicles of emotional arousal. The more frequently we switch to our smartphone or tablet, the less likely our emotional intensity falls, and the faster we recover states of higher emotional arousal. This helps explain why in study after study, we saw that the use of smartphones and media multitasking during traditional television viewing goes up significantly during commercial breaks. We tune into television for programming that is relevant and engaging, whereas advertising sometimes engages us but often is just boring.

Thanks to our digital devices, we need not tolerate the low arousal of boredom. We have all had the experience of reaching for our smartphones when we are just a little bit bored: standing in line at the store, in an elevator, on the bus, or in the car. The more often we switch to a new and more stimulating task, the more often the brain feels rewarded. But there is a downside. In fact, we are chasing our own tails. As a result of staving off boredom with digital devices, we become less tolerant of dull tasks and therefore get bored more easily. It is a vicious cycle. We have learned to manage boredom and other uncomfortable feelings by using mobile media, communications, and information technology as a type of mood regulator, even though, as we will see, these tools contribute to negative moods.

Good News and Bad News

The Time, Inc., study generated amazing insights into the drivers of media multitasking. At least in some cases, we are motivated as much by the potential pleasure and reward of emotional engagement as by the escape of low emotional arousal and boredom.

But the study left open important questions about shrinking media attention spans across generations. Are the brains of digital natives actually wired differently, thanks to the media they consume and the technology they use? Is differential attention span a product of different brains—brains changed by the world around us? Or are the divergences in media attention spans merely the result of preferences or physiological changes that track with age? In other words, if we repeated the study a few years later, would we see that the digital natives maintained their shorter attention spans even as they age, signifying a group effect? Or would we see that younger people have shorter attention spans, but as they get older, their focus improves as their brains mature, indicating an age effect? A group effect, also called a cohort effect, would mean that digital natives have shorter attention spans by virtue of some characteristics common to their group, characteristics that don't go away as they age—such as exposure to digital devices. An age effect would mean that there is no generational effect, that there is no specific brain difference between digital natives and digital immigrants.

In late 2016, almost exactly five years after the original study, Innerscope was acquired by the Nielsen Company. With access to more data and more resources, we had an opportunity to replicate the research with a larger sample.[16] We found evidence for both the generational group effect and the age effect: media attention span is a function both of growing up wired and of getting older. We interpreted the new results to be good news and bad news.

First the good news. Five years after the original study, with a new sample of both groups of participants in very similar study conditions, media attention spans showed the same trend as in the original 2011 study: older participants on average demonstrated a longer media attention span. This suggests that as our brains mature with age, we can focus longer.[17]

The bad news? In the new study, the oldest population showed a slight decrease in their attention spans compared to the oldest population in the earlier study. In addition, the difference in media attention spans between the older and younger groups in the new study was less than it had been in the original study, despite the fact that the age difference in the second study was greater. These are only two relatively small studies, so we must be careful not to weigh too heavily the findings, but they should still give us pause. The results suggest that as media technologies continue to penetrate our lives and media multitasking becomes more pervasive, there is a trend toward less tolerance for boredom and shrinking attention spans across all age groups. It appears that we are all being rewired to some degree.

It is also concerning that the media attention spans in both studies were remarkably short: the average media attention span we found across both studies was under three minutes. We can be confident that this is short because we can compare our results to studies of similar tasks. This is important when evaluating attention spans: the researcher must carefully consider the task in question because different tasks can be expected to command different amounts of attention. Whether we are studying for an exam, writing a report for work, paying our bills, or playing video games, measured attention spans vary greatly by task. In this case, the task is engaging with one media platform or device while another media platform or device is nearby.

Because media multitasking is a relatively new behavior and the research to capture it is expensive, it is hard to find comparisons to put our results into context. But research from 2008 offers a relevant benchmark. That study investigated the impact of background television on children's play and measured the media attention spans of one-year-old and three-year-old children as they played with toys while the television was on near them. Although this clearly is not exactly equivalent to the studies I undertook with Innerscope and Nielsen, the methods used to calculate media attention spans were very similar.

The 2008 study found that three-year-old children had an average attention span of 1.8 minutes.[18] When we compare this to our own results, we see that the digital natives' attention span is only 22 percent higher than that of a three-year-old, while the digital immigrants' attention span is 56 percent higher. This suggests that, roughly speaking, today's young adults have media attention spans significantly closer to those of three-year-olds than to those of adults. While not a perfect comparison—more research is certainly needed—the results suggest that the proliferation of media devices has increased media multitasking to such a degree that our collective media attention spans appear to make little progress well into adulthood.

What is the impact of shorter attention spans? Our 2011 Time study included an online survey of 1,700 adults nationwide, which found that digital natives reported that they were more easily bored, more easily distracted, and were more likely to get nervous, compared with digital immigrants. We cannot know for sure all the reasons for these gaps, but the survey also found that digital natives were significantly more likely than digital immigrants to carry their smartphone from room to room, to report that their smartphone was the first thing they used if they woke up in the middle of the night, and to indicate that they "prefer texting to talking to people."[19] This connection between emotional distress and dependence on digital tools was strengthened by a 2018 study in which US teens reported they would feel bored and anxious to a "moderate" or "great" extent if they had to go without their smartphones for a day.[20] These results foreshadow some of the mental health consequences of frequently using media as a mood regulator.

In the concluding section of our Time paper, we called for more sophisticated research methods and models, informed by more sophisticated brain and behavioral science, to stay abreast of how and why media is increasingly consumed at such high rates across platforms, cultures, and generations. Since then, researchers in diverse disciplines have been inspired by changes in media habits to learn more about how the brain itself is changing in response to these behavioral changes. The result is the beginning of a neuroscience-informed model of engagement that shows how new media, communications, and information technologies are altering us at a physiological level.

We are just beginning to understand the consequences of the massive shifts in media habits and consumption patterns that have occurred since the introduction of smartphones into our lives. But one thing is clear: change is the new constant. And thanks to lifelong neuroplasticity, our brains keep changing in response to new media and technologies regardless of our age.

The precise reasons that our brains are so responsive to media stimuli are not always easy to describe. We know the answers to some of the underlying questions; in other cases, we are left to speculate on the basis of available information. To the extent that there are answers to why our brains engage with smartphones and related technology in the ways they do, some of the answers are informed by a trip back in time. It is not enough to pay a visit to 2011, or the iPhone rollout, or the birth of the internet. We have to look to the early days of the human species, long before what we now think of as technology was a part of our existence.

2

THE POWER OF THE PREFRONTAL CORTEX

Imagine you've left the relative warmth of the southern hemisphere to travel hundreds of miles to the cooler northern latitudes in search of food. You're with a group of fellow early humans. You're tired, cold, and hungry. At some point in your journey you encounter a small group of shorter creatures that share many of your features. You don't understand their language or their customs.

These are probably Neanderthals or Denisovans, the first humanlike species to leave Africa and wind up in Europe and Asia, respectively. They lived there for hundreds of thousands of years before *Homo sapiens* also left Africa and eventually took over the world. What do we know about why our species survived and these early hominins didn't? There are many theories, but the key distinction—the source of modern humanity's adaptive advantage—likely comes down to the brain, and specifically the prefrontal cortex.

Neanderthal fossils offer a useful record of our earliest cohabitants because they are plentiful and because Neanderthals and humans shared the same spaces for tens of thousands of years. Physically, Neanderthals had shorter stature, barrel chests, and lighter skin. Their heads were also shaped differently, with larger jaws, bigger eyes, and more sloping foreheads. This shape accommodated a brain similar in size to humans' but different in structure. In particular, evolutionary theorists speculate that Neanderthals developed larger eye sockets as well as larger occipital and other brain areas associated with visual processing. Europe has longer nights and cloudier days than Africa, so it stands to reason that Neanderthals benefited from greater vision in low light.[1] But larger eyes and increased visual processing abilities came at a cost. Neanderthal brains were more accommodating in the rear, where visual processing occurs. But they were correspondingly smaller in the front part of the brain known as the prefrontal cortex.[2]

The prefrontal cortex, sitting just behind our foreheads and comparatively modest eye sockets, is the most highly evolved part of the modern human brain. It is hard to overstate the uniqueness of the prefrontal cortex relative to other brain regions. It accounts for nearly 35 percent of cortical gray matter and guzzles a disproportionate share of the calories we consume.[3] It is thought to be the most *wired* part of the human cortex, possessing a high density of interconnectivity with almost all other brain areas, especially the emotion and reward centers in the lower subcortex. Anatomically, the prefrontal cortex is in the privileged position of activating, suppressing, and otherwise coordinating a wide range of complex brain networks underlying a vast array of uniquely human skills. Indeed, the prefrontal cortex does so many things that it is difficult to fully capture its importance to cognition, emotion, and behavior.[4]

Among the key roles of the prefrontal cortex is to enable what psychologists and neuroscientists call executive function. This broad term encompasses a foundational part of our lives. Simply put, executive function is the ability to represent in one's mind multiple plans and execute them (or not) using a wide range of goal-directed thoughts and activities.[5] In this respect, the prefrontal cortex is the critical coordinator of our lives. The brain and all its complexity inspire many metaphors, but for the prefrontal cortex, the image of an orchestral conductor is apt.

The prefrontal cortex calls on and harmonizes various brain regions as needed depending on our environment—picking up the tempo of the brass section here, quieting the percussions there. In lieu of a baton, the prefrontal cortex uses electrical impulses and neurotransmitters to coordinate activity and communications crisscrossing different brain areas. This is how the prefrontal cortex controls attention, retrieves information from our emotion and reward centers, and accesses a treasure trove of past experiences in our memory centers. These experiences are our repertoire, guiding complex behaviors we manifest for survival, socialization, and success in a complex world.

With respect to attention, the conductor of our brain's symphony of neurons must turn on and off different elements with precision timing to make music and suppress noise. Our ability to focus depends on response inhibition (that is, impulse control or self-control). The inhibition enabled by the prefrontal cortex allows us to put the brakes on thoughts and behaviors that may be inappropriate or distracting and thereby keep attention on the task at hand. Sustained attention is of course critical for learning, memory, and thriving. It is also compromised, as we will see, by modern media and technology multitasking.

The brain's emotion centers interface with the orbital frontal portion of the prefrontal cortex, which was likely more developed in early humans than in other hominins. This area allows us to bind thoughts, memories, and experiences with their corresponding emotional states.[6] The experience of our own emotional states and the ability to read the emotional states of others enables us to form strong social bonds, which are critical for the complex ways in which humans collaborate and cooperate with one another. The prefrontal cortex helps us regulate and interpret negative emotions and underlies our empathic and other prosocial skills. It is active when we look at infants and people we care about, and as discussed in a later chapter, a larger prefrontal cortex has been correlated with larger social groups across species.[7] As we use mobile media devices to manage our emotions and regulate our moods, we are changing the nature of how we navigate our lives and relate to one another.

Finally, the relationship between the prefrontal cortex and our memory system is what allows us to take a sequence of discrete external events and integrate them into a coherent story, like the various musical themes that combine to form a symphony. But unlike a real-life conductor who typically directs one performance at a time, the prefrontal cortex has access to many scores stored in our memories through experience, which can be called upon at a moment's notice and played at any point. We can draw upon our memories in order to improvise as well, adapting to the changing environment. The ability to form new memories and call upon older ones underlies all manner of human knowledge-sharing, risk management, decision-making, and strategic thinking. We will see how access to copious amounts of information via the internet is changing the nature of how we store and use experiences and knowledge.

Neanderthals and other early hominins lacked the large, densely connected prefrontal cortex that early modern humans enjoyed.[8] Perhaps this was their downfall. Humans had superior focus, more complex emotional relationships, and more sophisticated use of memories, enabling us to learn from the past and plan for the future. As Robin Dunbar, an Oxford University researcher, put it, Neanderthals "were very, very smart, but not quite in the same league as *Homo sapiens*."[9]

Our ancestors' unique ability to engage with the world and focus sustained attention, manage emotions, recall and exchange information, and create strong social bonds might have been enough to "tip the balance when things were beginning to get tough at the end of the last ice age," Dunbar explains, referring to a critical time when the climate changed dramatically. Early modern humans used their larger prefrontal cortex to adapt. Neanderthals, Denisovans, and other early hominins could not survive.

Beyond the vision issue, another factor in the divergence of human and other hominin brains may have been the structure of early human societies. One theory holds that the warmer climate contributed to larger populations in Africa that encouraged strong social bonds and facilitated longer relationships and coparenting. This allowed for longer periods of nurturing and more careful development of the fragile young brains of early human infants, offering a competitive advantage over time.

This is known as the social brain hypothesis, and it suggests that there is a neurobiological basis for the special social wiring of the modern human brain—wiring so powerful that it may have changed the course of our existence on the planet. The social brain hypothesis essentially states that our larger prefrontal cortices evolved under the intense social competition produced by early population density. Increased social competition resulted in increasingly sophisticated social structures, including the notion of social status, related to reproductive success.[10] Strong social bonds and the formation of complex social networks allowed for more active child-rearing over longer periods of time. More complex social structures also led to more knowledge-sharing, which aided the invention of more sophisticated tools. As human life expectancy increased slowly over time, our technologies improved, as did our dominance of and influence on the planet. And after a fair amount of interbreeding with humans, Neanderthals and Denisovans disappeared forever.

The human brain, with its impressive prefrontal cortex, the conductor of neuronal symphonies and master of the orchestra that is the rest of the brain, is the difference between impulse and insight, distraction and focus, reaction and reflection. The prefrontal cortex does not generate emotions, but it is critical in interpreting our emotional world. It is the difference between anger and empathy, social pain and social comfort. It is critical to our ability to learn new information and call upon prior memories. And as we will see, it plays a role in the difference between habit and addiction.

As remarkable as the human brain is, there are real questions about how it will manage the technologies we are creating and using in the digital age. The prefrontal cortex, our most precious brain resource, is under intense pressure in our current age of distraction, as we risk turning our personal symphonies from harmony to cacophony.

Early Experiences, Lasting Impact

By the early 1980s, the Romanian dictator Nicolae Ceaușescu had a serious problem. He had bet heavily on a national economic strategy that involved building massive oil refineries with billions of dollars borrowed from Western countries. But his productivity projections were too rosy, and soon construction delays were piling up. There were other bad investments, too, and the state was struggling to pay for the cleanup from a massive and unexpected earthquake. The debt became crushing.

In an attempt to make good on the loans, Ceaușescu initiated a series of nationwide austerity measures and ordered the export of most of the country's industrial and agricultural products. This resulted in massive shortages of food, heating oil, and electricity at home, devastating the Romanian economy and the people's quality of life. Making matters worse, the population had blossomed in the two decades preceding. Back in the 1960s, Ceaușescu had tried to stimulate economic growth by encouraging demographic expansion, which he pursued by making contraception and abortion illegal. Shortages in household goods collided with a population explosion, exacerbating the economic collapse.

Unable to afford food for themselves, let alone their children, Romanian parents reluctantly turned over tens of thousands of young kids to a burgeoning network of underfunded orphanages. Overcrowded and under-resourced, the orphanages struggled to provide necessities like heat and nutrition. Reports of severe neglect, unsanitary conditions, and rows and rows of cribs with crying babies began leaking to the West. The debt was finally repaid in 1989, but by then the Romanians were in revolt and Ceaușescu was forced to flee angry mobs by helicopter. Soon after, he and his wife, also a high-ranking politician, were bound together and executed by a firing squad.

A year later, the television news program *20/20* aired the first video of the shocking conditions experienced by nearly a hundred thousand children.

Western aid organizations and psychologists poured into the country to help. But some researchers saw an opportunity to do more than provide immediate assistance. They also wanted to take advantage of the unfortunate circumstances to answer an important question: What is the impact on young developing brains of such extreme neglect in early childhood?[11]

We know that early childhood is a critical period of brain development. This is intuitively clear, and there is ample data to back up what experience tells us. Modern developmental neuroscience has shown the astonishing speed with which the young brain grows. Researchers at the Harvard Center on the Developing Child estimate that, in the first few years of life, millions of new neuronal connections are formed every minute, explaining how young children go through developmental milestones so quickly.[12]

As the milestone metaphor suggests, these developmental stages come in order. Every one of us follows the same path, and healthy infants will achieve the various early milestones at roughly the same age. In the initial months of life, the infant will develop sensory pathways for sight and sound, toward the back of the brain. This is followed by the development of brain areas behind the prefrontal cortex that control large and small movements facilitating gross and fine motor skills, eventually allowing the infant to sit up, point at objects, and begin exploring the world. Toward the end of the first year, language processing shows signs of progress with utterances and a few early words, as do some higher cognitive functions including other communication skills, primitive spatial relations, and object recognition.

The neuronal blossoming and development continues at breakneck speed until about the age of four, with the cerebral cortex especially gaining in volume, thickness, and surface area. Thereafter growth continues, but more slowly. By six years of age, the average human brain is four times larger than it had been at birth and about 90 percent of its adult volume.[13] But that is not the end of the story. We now know that the human brain is quite dynamic, forming new neuronal connections and other structural changes throughout the lifespan, right up to the point of death.[14] And while many areas of the brain have reached near maturity by the early school years, one area that continues to show signs of extensive change thereafter is the prefrontal cortex. It will be another two decades before that critical brain region is fully grown.

Thus modern neuroscience teaches us that brain development isn't limited to childhood. Indeed, we are constantly rewiring ourselves. But brain development is also a fragile process that is easily influenced by environmental factors large and small. And the human brain is especially delicate in early development. Someone needs to nurture, feed, clean, and protect us long enough for our brains to mature so we can live independently and, in turn, nurture someone else. In other words, strong social bonds are critical to healthy brain development.

This is where the mass neglect of Romanian orphans comes in. If early childhood brain development is fragile, and if we need the attention of adult caregivers both to provide for basic needs and to build the foundation of our social lives, what is the impact of the caregiving relationship being interrupted, damaged, or nonexistent? What happens, in other words, when children lack the continuous social attachment to a community of caring providers?

The Importance of Early Attachment

One of the leaders advancing our understanding of the developing brain, social relationships, and related neurobiology is Dan Siegel, a psychiatrist at UCLA. I was introduced to Siegel through his 1999 book *The Developing Mind* and later got to know him personally.[15] In the book, he lays out a neurobiology of social connection that begins at birth with parents and caregivers and continues with friends, lovers, and family until our final breaths of life. Drawing on attachment theory developed by the British psychoanalyst John Bowlby, Siegel's work shows us that the impulse toward social attachment, particularly early attachment, is encoded in our biology. We are instinctually motivated to form special social bonds, and to feel these bonds strongly.

Following the work of Sigmund Freud, Bowlby and others in the 1960s began to hypothesize that childhood relationships with parents and caregivers shape many of our behaviors and our ability to form the social bonds that are the basis of close relationships later in life. This is the core of attachment theory. At its basic level, attachment theory posits that reliable, consistent, emotionally responsive, face-to-face caregiving enables children to develop a secure style of attachment characterized by adept emotional regulation, a willingness to depend on others, and a high degree of comfort with close relationships. All of these are valuable skills in the pursuit of social bonds across the lifespan.

On the other end of the spectrum, inconsistent, emotionally distant, or unresponsive caregiving fosters a more insecure style of attachment that manifests as ongoing anxiety about abandonment, poor emotion regulation, intolerance of social intimacy, and avoidance of close relationships. Attachment theory posits that children internalize early relationship experiences and that these experiences inform their future memories and beliefs about social bonding. In turn, these memories and beliefs influence their brains' ongoing ability to regulate emotions, engage in healthy exploration of the world, and relate to others.[16]

The poor attachments and weak social bonds that result from inadequate early caregiving can disrupt our ability to process what Siegel calls the "flow of information" between people. This flow is critical to forming lasting connections and to healthy development. Major disruptors in this flow of social information are often referred to as adverse life events and include the loss of a parent, divorce, life-threatening illness, parental mental illness, and physical or sexual abuse. These adverse events can have a lifelong impact regardless of the developmental stage during which they occur.

That, at least, is the theory: our early social bonding has a powerful influence on our brains and on how we relate to others later in life. But can we prove it? Multiple researchers have tried to test the impact of early socialization by exploring how these experiences alter neuronal pathways in the brain that are thought to be implicated in the formation of strong social bonds later in life.

Consider two studies illustrating the evidence that early attachment influences us profoundly as adults. The first, published in 2000, followed a group of infants from white, middle-class families over a twenty-year period. The researchers sought to examine the stability of the participants' attachment styles over time. Well-validated assessment tools were used to measure participants' attachment quality and style as infants and again, two decades later, as young adults. More than 70 percent of the participants received the same attachment classification in early adulthood as they did as infants. In cases where the classification changed, adverse life events were a major contributing factor, supporting attachment theory's central claims.[17]

The second study comes from Robert Waldinger at Massachusetts General Hospital and the Harvard Study of Adult Development, one of the longest continuous longitudinal studies in history. Waldinger's attachment research focuses on eighty-one male participants across the lifespan.[18] The men recruited were from two sources. About half the group were from disadvantaged families in low-income Boston neighborhoods. The other half were Harvard sophomores. The study began in 1938, and for four years, psychiatrists and social workers conducted lengthy interviews with each participant, while independent observers watched and listened. Over the course of more than ten hours of conversation, the participants talked about their relationships with their parents. The observers, who were unaware of the goals of the study, coded the quality of each participant's relationship with his mother and father on a 5-point scale from distant, hostile, or overly punitive relationships (1) to nurturing relationships that encourage autonomy and foster self-esteem (5).

Two decades later, the same study participants engaged in two-hour follow-up interviews at midlife (ages 45–50). Again, independent raters observed the interviews. This time the raters evaluated participants' adaptive styles—that is, the ways in which they coped with difficult times in their lives—and their ability to regulate emotions during challenging relationship moments, stressful periods of work, or when dealing with health issues. Finally, surviving participants did an additional forty-five-minute interview years later, after age eighty. The interview focused on their social bonds with current significant others, including communication style and expressed levels of comfort, satisfaction, intimacy, and love.

Assessing the decades of data, Waldinger and his coauthors found that individuals raised in nurturing environments are more likely to have healthy relationships as adults. The authors conclude, "Warm, nurturing parenting fosters adaptive emotion-regulatory styles that facilitate effective relationship engagement later in life." However, once again, the research showed that the effects of early-childhood attachment are not immutable. Warm, nurturing parenting that leads to secure social bonds can be undone by negative life events.

To further bolster the case for the importance of early-childhood bonding in the formation of attachment style, let us return to children raised in institutions such as the Romanian orphanages. Researchers have found over the years that children who experienced neglect under Ceaușescu's regime have reduced IQs, difficulty with social functioning, higher rates of attention-deficit hyperactivity disorder (ADHD), and higher rates of forms of psychopathology relative to noninstitutionalized children.[19] The institutionalized children also have reduced brain volume (based on brain scans) and lower-quality brain activity (based on EEG), when compared with Romanian children assigned to foster care, in which foster parents were paid to care for the children in their homes.[20]

These and other results sadly but strongly suggest that early influences on brain development have lasting impact. "We can now say with confidence that the psychosocial environment has a material impact on the way the human brain develops," said Joan Luby, another researcher of early negative childhood experiences on brain development and a psychiatrist at the Washington

University School of Medicine in St. Louis. "It puts a very strong wind behind the sail of the idea that early nurturing of children positively affects their development."[21]

The results of these various studies vindicate the theories of Freud, Bowlby, and others who promoted the idea that healthy early-childhood social bonds are important for our future mental health and relationships. Qualitative and quantitative research, using techniques as diverse as coded interviewing, long-term cohort studies, and brain scanning, consistently finds that early-childhood socialization, along with emotional support from caregivers, has lifelong impact on social and emotional health. The question I propose is whether today's always-on digital culture undermines the early-childhood influences that we know are critical for our emotional and social health later in life.

The Changing Nature of Parenting

New mobile media, communications, and information technologies, combined with expanding access to high-speed internet, have ushered in one of the most significant periods of change in child development and parenting in the last hundred years. Today children are no longer exposed to just a few close relatives or other adult caregivers and their ideas. Now children, from a very early age, are routinely exposed to the attitudes, beliefs, and values of a wide variety of voices through an ever-growing volume of media content on an ever-increasing number of screens. Meanwhile parents and other adult caregivers—whose attention and emotional attunement to their young children is so important—are more distracted than ever, thanks to the same technologies.[22]

How much media are children exposed to today? Despite repeated warnings from the American Academy of Pediatrics (AAP), the average age of "regular" media consumption—meaning repeated exposure—has dropped dramatically in recent decades. In the early 1970s, the average age of regular video consumption via television was approximately four years old. Current estimates put the average age of first exposure at four months, with regular exposure starting at nine months of age.[23] And these very young children are not just watching a few videos. Surveys suggest that they are consuming as much as three hours per day. Given that nine-month-old children are awake for only ten to twelve hours a day, this represents an alarming proportion of their waking time.[24]

Children's media habits today go beyond traditional television. In 2015 children between the ages of two and eleven years old were spending on average about 60 percent of their media-consumption time watching TV. That means that 40 percent of their media-consumption time was spent using an internet-connected device, typically to watch downloaded or streaming video from Amazon, Netflix, and YouTube.[25] Another study from the same year asked parents if their kids, aged zero to four years old, had used mobile devices. Ninety-five percent of the parents, who were recruited from a pediatric clinic, said yes. Seventy-five percent of young children already owned their own device, and most two-year-olds in the same sample used mobile devices daily.[26]

No one thinks that a steady diet of media content in early childhood is akin to growing up neglected and malnourished in an orphanage. But there are reasons to worry that the high-tech lifestyles of children and adults mean that young kids are getting less of the attention and attachment they need to develop into grown-ups who have healthy relationships. Understanding the impact of this new type of stress is important and worthy of exploration. That will come in part 2, where I look at the consequences of excessive media exposure in early childhood and beyond. But first, what about the adult social brain? What does that look like, and how might it be at risk in the digital age?

3

WIRED TO CONNECT

When I was in my third year at Harvard Medical School, I was assigned to an outpatient family medicine practice north of Boston for two months. Medical school rotations traditionally had been one month long, but back then there was a renewed emphasis on primary care, and changes in the curriculum had medical students spending more time in outpatient clinics to understand how health issues evolved outside the hospital setting. I quickly understood that I was very fortunate in my assignment to the solo practice of Dr. John Abramson.

Abramson had a unique way with patients. Calm in his demeanor, he was able to connect with them in a way no other Harvard clinician in my experience did. As we got to know each other, I learned that he had supplemented his medical training by working with a shaman in South America and by reading extensively in diverse philosophical and religious traditions. These experiences informed his approach to medical care and inspired me to think more broadly about the special social bond that is the patient-doctor relationship.

It helped that at the time, there was growing evidence from placebo and psychotherapy research showing that the clinical relationship had healing power independent of medication, surgery, and physical touch. Later, as a psychiatrist in training, I learned this firsthand by doing psychotherapy. My experience, and the growing medical literature, showed that the patient-therapist relationship is often as powerful as Prozac and other medications in terms of emotional healing and clinical outcomes. The therapeutic bond, in other words, has the potential to be transformative and healing. How is this so—why are humans capable of forming such bonds?

I became fascinated with the idea that we are all wired to connect, and that much of the underlying brain functioning related to social bonding plays out below our conscious awareness. I was determined to measure this unique connection. I convinced Abramson to fund pilot research that would use basic physiologic sensors to probe the nervous system's involvement in social bonding. I wired him and a few of his patients to two skin-conductance devices, which allow real-time measurement of physiological arousal, and I borrowed my parents' video camera to record them during patient visits.

By watching the time-locked videos and biometric responses of patients and this gifted clinician, I was able to capture a type of physiological synchrony: during shared moments of meaning, both the patient and Abramson exhibited similar physiological responses. I later applied the technique to measure the experience of empathy during sessions with more patients and other clinicians. I found that the more often these moments of physiologic synchrony occurred, the higher patients rated their therapists' empathic abilities. The capacity to mirror someone else physiologically was related to our experience of empathy and feeling understood.[1]

As I was collecting this data, I joined other researchers who were doing neuroimaging studies and taking direct measures of brain activity during psychosurgery. The results showed that physiologic responses, like the skin-conductance responses I was testing, were related to stimulation and activity in the prefrontal cortex and emotion centers of the brain. The neurobiological link between cognition, emotion, and arousal in the prefrontal cortex was becoming increasingly well established. Indeed, we were starting to map those links with considerable precision.[2]

As I moved away from academic research on the patient-clinician connection and into the world of consumer connections with media and marketing content, other scientists continued to explore human social interactions using a variety of techniques to measure interpersonal physiology—the synchrony of physiological responses in individuals as they communicate and relate to each other. Importantly, these studies included direct measures of associated brain activity.[3]

Research outside of clinical care suggested that the biological basis of physiological synchrony is directly related to the formation of social relationships. In one example, researchers in Spain used simultaneous EEG measures of brainwave activity in people having casual, face-to-face conversations. The results showed moments of brainwave synchrony very much like the physiologic patterns I saw in patient-clinician pairs. As one journalist covering the research put it, there is "an interbrain communion that goes beyond language itself and may constitute a key factor in interpersonal relations." The neuronal synchrony was so pronounced during face-to-face social interactions that researchers were able to determine when the two people were in conversation based solely on analyses of their brainwaves.[4]

Neuroscientist Morten Kringelbach, who uses brain-imaging technology to study the mechanisms of parenting instincts, relates interpersonal physiology to our "need to nurture." In one study, Kringelbach asked adult participants to look at photos of infants or other adults while the participants underwent brain scans. He found activity in the orbital frontal cortex, a part of the prefrontal cortex. This brain region, which is highly connected to the brain's emotional centers, was active within milliseconds of exposure to images of infants; responses to images of adults came more slowly. Below our conscious awareness, our brains are automatically working to foster bonds with the very young.[5]

Further evidence of the physiological basis of social bonding comes to us from the social psychologist and cognitive neuroscientist Matthew Lieberman. His work offers tantalizing clues about the relationship between physical pain and the emotional pain that follows social rejection. In a now-classic experiment, he had participants lie in a brain scanner and play a video simulation that mimics a game of catch. In reality the participants were playing against the computer, but they were led to believe they were actually playing with two other research participants. During the first half of the experiment, when the participant waves her hand, the computer has the "other person" throw the ball to her. In the second half, the game changes so that regardless of what the participant does, the other players never throw her the ball. Brain scans of the participants showed that some of the same areas of the brain that are active during experiences of physical pain are active during the perceived experience of social rejection.[6]

Lieberman and others argue that this relationship between physical and social pain reflects the necessity of relationships for our survival. We are wired to connect because, if we did not experience the care and support of others, we would be unable to survive infancy. Our vulnerability in the first days of life demands the care of others; the pain response caused by the absence of others reflects our instinct toward togetherness, lest we suffer terrible consequences.

Pain responses evolved to signal us, to point toward problems that need to be addressed or situations to be avoided. This is as much true of the pain we experience when we hurt our bodies as of the pain we experience when social bonds are broken. After all, from the perspective of the brain, these are much the same thing.

The Changing Nature of Social Connection

Our need to connect has combined with the availability of high-speed, always-on devices to produce a social media explosion that dramatically shifts the ways we interact with each other. The rise of smartphone apps that take advantage of broadband connectivity and sophisticated, miniaturized camera technology has made it incredibly easy to share ideas and experiences with family, friends, and strangers the world over. Social media has truly taken over how we connect online—according to Nielsen, social media surpassed checking email as the most common reason for logging onto a computer device back in 2014. Other statistics show that in a single day, Facebook users share 2.5 billion pieces of content, YouTube users upload 12 years' worth of video, Instagram users post 40 million photos, and there are 400 million posts on Twitter. In a single day![7]

Social media facilitates some of the same exercises that are critical to the formation of social bonds: sharing stories about ourselves and learning about the lives and experiences of others. Research shows that nearly 40 percent of everyday speech is used to relay information about our intimate experiences and our social lives. Social media studies suggest that the number of posts related to some aspect of our communal living (as opposed to news, ideas, or opinions) are even higher.[8]

What matters here is the act of sharing information with others. It's not enough to think out loud while no one else is listening. One brain-scanning study has shown that when we disclose our inner thoughts to others, the brain releases dopamine, much as it does when we eat chocolate, take drugs, or have sex. The likelihood of dopamine release is higher when we disclose these thoughts to another person than when we are just thinking to ourselves. Pondering these findings, the study's lead scientist said, "I think this helps explain why Twitter exists and why Facebook is so popular, because people enjoy sharing information about each other."[9]

By far, social media's greatest impact has been felt by teenagers. Teenagers, who are at a unique developmental stage, increasingly rely on smartphones and other internet-enabled devices with access to Facebook, YouTube, Twitch, Tumblr, Snapchat, Reddit, WhatsApp, Kik, Instagram, and other apps for social connection. One survey comparing teen and parental usage of social media across the United States, Britain, and Japan found that, in each country, half or more of teens report checking their devices at least hourly. About the same proportion of teens in all three countries felt "addicted" to their devices. Interestingly, a quarter or more of teens in each country thought their parents were addicted to a digital device as well.[10]

I discuss social media's role in the lives of teenagers in more detail later in the book—how their development is affected by social media use and how to draw the line between addiction and relatively innocuous habits. But it is worth keeping in mind that teens are not the only group increasingly dependent on social media. Perhaps surprisingly, the heaviest user group according to 2016 data were adults ages 35–49, who spent almost seven hours per week on social media. That study found the greatest use was by Generation-X adult women on Sunday evenings during prime-time hours, which suggests a high degree of media multitasking.[11]

Among the many social media networks, by far the one most accessed globally via smartphones is Facebook. In 2013 Facebook touted a major milestone: it had accrued more than a billion monthly users in less than ten years. In early 2017, just a few years later, the number had doubled. Not only are more people using Facebook, but the milestone of 2 billion monthly users included a second record for Facebook: 66 percent of users were taking advantage of the platform daily, up from 55 percent in 2013. Much of the more recent increase in Facebook users comes from the developing world and the Asia-Pacific region, so that today Facebook is truly global: more users are accessing the app more frequently across a larger portion of the planet. By the second quarter of 2021, the platform had 2.9 billion users.[12]

Facebook's triumph was fueled by its well-timed shift to mobile devices. Now its users—hungry for sharing ideas and connecting online and increasingly depending on the platform for news and information—can be reached in one easy-to-access location. Advertisers took notice, as Facebook offered an appealing new way to reach this mass audience via hypertargeting of individuals at the cost of pennies each. The result was huge revenues for the company. Mobile ad revenue accounted for the bulk of Facebook's $27.6 billion haul in 2016, up 54 percent year-over-year. In 2017 and 2018, the hot streak continued with even higher quarterly sales and annual revenue. By 2020, total revenue topped $86 billion. All driven by advertising.[13]

Despite such outsized growth and financial success, Facebook faces some very real challenges. Its sheer size and influence on our minds and moods, discussed in more detail in part 2, is starting to challenge its own business model. And scandals concerning data privacy, foreign influence, and fake news are testing our notions of what is appropriate to share online. The number of concerning reports is growing, as Facebook proves unable or unwilling to prevent the dissemination of false information on its platform and, in some cases, appears to encourage it. Russian hackers famously took over accounts and spread disinformation to influence the 2016 US presidential election; advertisers were allowed to target "Jew Haters"; depressed and desperate users were allowed to livestream suicides; and others were able to share images of child sexual abuse.[14] In 2021, after the *Wall Street Journal* revealed that Facebook's own research showed its subsidiary Instagram has a harmful impact on the mental health of teenage girls, the company was reduced to bombarding user newsfeeds with pro-Facebook stories in an effort at damage control. "They're realizing that no one else is going to come to their defense, so they need to do it and say it themselves," said Katie Harbath, a former Facebook policy director.[15]

All this negative feedback is leading other past Facebook executives to do some soul searching. Chamath Palihapitiya joined Facebook in 2007, just three years after its launch out of a Harvard dorm room. His job as vice president of user growth was to figure out how to get more people on the platform. In a 2017 forum at Stanford University, he admitted having "tremendous guilt" about the company he helped to form. "The short-term, dopamine-driven feedback loops we've created are destroying how society works," Palihapitiya lamented. "No civil discourse, no cooperation, misinformation, mistruth. And it's not just an American problem . . . this is a global problem." Palihapitiya is not concerned about Facebook alone but the online ecosystem generally:

> We curate our lives around this perceived sense of perfection because we get rewarded with these short-term signals, hearts, likes, thumbs up, and we conflate that with value and we conflate it with truth, and instead what it really is, is fake, brittle popularity, that's short-term and leaves you even more . . . vacant and empty than before you did it . . .

and think of that compounded by 2 billion people. . . . You don't realize it, but you are being programmed.

Others, such as former Facebook product manager Antonio Garcia-Martinez, say that Facebook and Silicon Valley at large "are out of control."[16]

With great size and great revenue come great responsibility—at least that is what Facebook executives say they believe. Senior executives, including Mark Zuckerberg, have been going on "listening tours" around the world, while the company is putting millions of dollars into buying television advertising in hopes of burnishing its image. Facebook now employs Washington lobbyists and reports mea culpas to Congress and its users to help stave off calls for regulation. Indeed, there has been a degree of voluntary reform at Facebook in the wake of scandal and controversy, but will it be enough? It is sometimes hard to know whether internal changes are substantive.[17]

The speed of adoption and the power of social media and other mobile communication and information technologies, along with their near ubiquitous connectivity, hyperinteractivity, and broad array of functionality should inspire awe. These technologies empower us in unimaginable ways. But the same technologies should also cause us to pause, reflect, and consider the consequences of all these changes. Those consequences include the rapid dissemination of misinformation and the solidification of new platforms for unhealthy social comparisons, hate speech, bullying, and illegal activity. Social media has hijacked our brains for better and for worse.

Recall that the reward centers in the brain are stimulated when we share autobiographical information. The power of social sharing that's wired in us has catalyzed much of the growth of social media over the last decade. But do the relationships we create online through social media foster or degrade our offline relationships—the ones that really matter for our thriving? If strong social bonds are dependent on a healthy prefrontal cortex and healthy brain, we can reasonably speculate that technologies that interfere with information processing degrade this critical brain region.

Strong social bonds have been forged between humans for eons through face-to-face communications and physical proximity. Healthy social relationships are critical to the development of our self-identity, our collective ability to form large social groups, and our capacity to achieve reproductive and technological success. Social bonds are necessary for our survival on the planet. Social media powerfully changes how we share information and how we relate to one another. I will make the case that it interferes with and rewires our reward centers and strains our prefrontal cortex with deleterious consequences. That is because our brains have very real limitations over which endless online content runs roughshod.

Limits of the Social Brain

In 1992 the British anthropologist Robin Dunbar published a highly influential paper whose provocative thesis is reflected in its title: "Neocortex Size as a Constraint on Group Size in Primates." In the article, Dunbar argues that there is a species-specific upper bound on the size of social groups, caused by limits on the information-processing capacity of the brain. The finite number of neurons in the cerebral cortex limits a species' ability to process information in general; of particular interest here, the limits of the cerebral cortex also constrain the ability to monitor and manage social relations. As a result, social networks can grow only so large. The theory suggests that if a social group's size exceeds the species' limit, the group becomes unstable and fragmented.

Dunbar offers proof of his thesis by showing a relationship between the size of a species' average social group and a key brain metric across a variety of nonhuman primates. The brain metric involves a simple ratio of cortical volume to the volume of the rest of the brain. Extrapolating results from nonhuman primates to human social groups and reviewing evidence for the average group size of hunter-gatherer societies over time, Dunbar came up with an upper limit for the size of human social groups, putting the number at around 150 social bonds. This figure, which is based on historical, statistical, and neuroanatomical data, is now commonly referred to as Dunbar's number.

Dunbar argues that human social-group sizes historically fell into three distinct classes: relatively small groups of between 30 and 50 individuals, often tightly clustered in time and space; a large population group of 500 to 2,500 individuals sharing geographic boundaries, language, rituals, and other culturally specific commonalities; and an intermediate group of 100–200 people, which "constitutes a subset of the population that interacts on a sufficiently regular basis to have strong bonds based on direct personal knowledge." In other words, there is something special about the people in this intermediate group, and it takes time and effort to form and maintain these relationships.[18]

The intermediate group size occurs frequently across a wide range of modern human societies. For example, since early Roman times, infantry units have consistently numbered between 100 and 200 individuals. Academic specialty groups tend to conform to this size. Some churches are enormous, counting thousands, even tens of thousands of members, but most congregations have 100–200 members. And modern analyses of staff turnover and absenteeism in businesses show that 150 employees is a critical limit for effective communication and management by a small group of people. Beyond this number, organizations often must create new internal structures with additional layers of management to function optimally.[19]

While initially well received, Dunbar's eponymous number was not without critics. But to his credit, he did not stop with his early findings. He went on to use brain-imaging tools to test his version of the social brain hypothesis more directly. In 2010 Dunbar and colleagues looked at forty healthy adult human brains and an assessment of a key aspect of social cognition called theory of mind. Theory of mind is an extension of empathic skills. It reflects a person's ability to attribute mental states, such as attitudes, beliefs, and emotions, to others and to understand that those mental states are separate from one's own. If you can accurately assess what another person believes about you, regardless of whether you also hold that belief, then you have a solid theory of mind. This is thought to be important for the formation of relationships and the maintenance of strong social bonds over time.[20]

What Dunbar and his colleagues found is that the size of the orbital frontal cortex, a region of the prefrontal cortex heavily involved in emotion regulation and social cognition, is highly correlated with scores on a psychological test that measures a person's theory of mind. Importantly, the same scores do not predict the volume of other areas of the prefrontal cortex, which have less involvement in emotion regulation and social cognition. The findings also support previous research showing that damage to the orbital frontal cortex results in a variety of social problems.[21] Subsequent research by Dunbar and colleagues has found that the volume of the orbital frontal cortex of each individual is correlated with the size of that person's social group and is mediated by social-cognitive skill as measured by the theory-of-mind task. Thus there appears to be credible evidence for a relationship among the volume of this subregion of the prefrontal cortex, social cognition, and the size of human social networks.[22]

Within our social networks, we tend to have an inner circle of about three to five people. This core group occupies about half of our socialization time.

The rest of our time is divided across the Dunbar number of 150 (in reality, his number is a range from 100 to 200, depending on one's social skills and brain capacity). This suggests that the quality of relationships tracks the amount of time invested in them.

Not surprisingly, membership in the inner circle changes over time. For example, when one enters a serious romantic relationship, one often commits a disproportionate amount of time to the romantic partner and sacrifices other inner-circle relationships. The same often happens when one has children: given a child's intense needs, parents often shift time away from other close relationships. There are only so many hours in the day, and our brains have limited capacity to socialize and connect.

The limit on our brains' ability to form social bonds challenges the utility of large online social networks. When asked about the implications of using more and more brain resources creating online social networks, Dunbar responds matter of factly, "All that seems to be happening when people add more than 150 friends on Facebook is that they simply dip into these normal higher layers." In other words, our so-called friends are really part of a larger, nonintimate social group beyond Dunbar's number. These contacts are not people we can meaningfully keep track of or have a relationship with. Many of our online friends aren't really friends at all.[23]

Fair enough; one might object that this isn't in and of itself a problem. Humans have always had a variety of strong and weak relationships, so there should be little risk of expanding our online social networks. But doing so comes at great expense in terms of our time and attention. Rather than use our time and our limited brain capacity to build high-quality social connections in the real world, we increasingly use these resources on the less meaningful online connections encouraged by our habits around mobile media, communications, and information technology.

If early relationships are critically important to our future relationships and our health, and if we have limited capacity for building relationships, how concerned should we be about the growing influence of these technologies on the quality of our social connections? What are the consequences of spending time with an increasing number of virtual "friends" rather than investing in real-world bonds? In phase two of our journey, I try to answer these questions. It turns out that brain and behavioral science have a lot to tell us about the radical changes and negative consequences our technology habits have wrought when it comes to the ways in which we process information and relate to one another.

PART II

REWIRED

Assaulted Brains

4

EARLY CHILDHOOD, INTERRUPTED

Ryan was four years old when he started unboxing toys on camera in 2015. A few years later—assisted by his mother, who left her job as a high school teacher to work with him full time—he was YouTube's highest-earning star, of any age. He made a record-breaking $29.5 million in 2020, releasing daily videos that blur the line between entertainment and commerce. His channel, now called *Ryan's World*, has generated more than 60 billion views. I found out about Ryan when my then-four-year-old son became enamored with the videos. As Ryan opened boxes and talked about new toys, he created a virtual space of limitless wonder for kids—including my own.[1]

When it comes to children's experiences, today's media ecosystem is dramatically different from the one in which I grew up. In addition to multiple cable television networks with ad-supported children's programming available around the clock, a proliferation of new online channels feature short, on-demand, "snackable" forms of amateur video. Plus, parents can download a variety of content from sources like Netflix and Amazon and play them on portable devices for their kids anytime and anywhere. In a sea of choice, young users move freely from channel to channel and device to device, from large screens to small ones, through an endless variety of professionally and not-so-professionally produced short-, medium-, and long-form content.

The shift in content creation, from professional producers of television and film to amateurs online, is important. The popularity of so-called user-generated content on YouTube, TikTok, and other outlets is fueled by computer algorithms that curate and serve up videos tailored to the individual tastes of young users. These algorithms are so sophisticated that in some cases they reveal—arguably create—previously unknown markets, markets that producers and advertisers are eager to reach to capitalize on children's surprising and peculiar preferences. Who could have predicted that Ryan's toy reviews and unboxing videos would take off? The unboxing genre did not even exist before a few years ago.

These changes raise many questions. What is the impact of shifts in media consumption by young children—increasing the quantity of content consumed, from new and questionable sources, via more screens than ever before? How is this sort of media consumption affecting the developing brain, and what are the downstream consequences for our physical and mental health?

As I've noted before, there is not always fresh, directly relevant research available to answer such questions. The media landscape is rapidly evolving, and science is catching up. But we can still gain insights by extrapolating from prior research on traditional media and integrating those findings with the growing understanding of childhood brain development provided by neuroscience. As I detail in this chapter, the available research shows that our brains, and our children's brains, are being rewired in ways that need to be examined. Before we proceed, however, it is important to raise some caveats, as there are several challenges inherent in interpreting decades of research on television and applying it to today's media world.

First, as noted, production and access happen differently in the new-media environment. In the not so distant past, video content was created and distributed by professionals and consumed at a particular time of day. Broadcasters set schedules that enabled parents and caregivers to limit children's viewing to professionally produced content like *Sesame Street*. First aired in 1969, *Sesame Street* was intended to provide high-quality, research-driven learning content for children, in particular kids living in poverty. Such programming required scores of writers, producers, directors, and actors. The show's success is undisputed. By contrast, today's user-generated amateur content, often made with less admirable goals, is produced by small groups and individuals wherever their smartphones take them. Past studies of media consumption assumed a different kind of media content than adults and kids consume today.[2]

Second, availability and presentation have changed. This often-amateur content is accessible 24 / 7 on mobile devices that parents often have trouble monitoring (more on that below). Today our children frequently consume media unsupervised, and can watch what they want, when and where they want it. Screen sizes have also changed; TV screens have gotten bigger, but most screen time is spent with small, portable images thanks to smartphones and tablet computers.

Third, the way we engage with new media is not like the way we engage with old media. Traditional television offers a passive experience: someone else produces and directs an immersive audiovisual story that unfolds linearly, with a clear beginning, middle, and end. We follow along. This experience triggers what neuroscientists call bottom-up processing, which refers to the lower or subcortical networks of neurons in the brain driving the process. This is very different from how we tend to engage with new media, particularly online or in apps, which are by nature more interactive and encourage more switching and skipping. New media therefore trigger more top-down processing, with the networks in the upper prefrontal cortex of the brain in charge.

Users engaged in top-down processing become the coproducers of their experience, making choices and reacting continually over time. As young users shift their attention styles and behaviors with new media, they tend to consume more media in shorter bursts, obtaining more immediate emotional rewards. As we will see, these attentional habits and emotional experiences leave their mark on the underdeveloped prefrontal cortex that mediates them, potentially contributing to shorter attention spans and other consequences later in life.[3]

Fourth, a major contributor to shorter media attention spans is media multitasking, defined as the use of two or more media devices simultaneously or the use of one media device in conjunction with a separate task such as homework. Given the ubiquity and mobility of modern media, the rise of media multitasking should not surprise us. The massive rise in media multitasking has implications for our children and ourselves. It also challenges researchers to think differently about the work they do. Older research had no reason to account for media multitasking, making this a new area that we need to examine from multiple angles.

Fifth, because small, mobile screens make it hard to tell exactly what our children are consuming, a lot of current research depends on questionnaires directed at parents and young users. This increases the likelihood that usage data will be subject to various kinds of bias, as people tend to both over- and underreport media use depending on how and when questions are asked. So it is important for researchers to integrate survey results with new methods to ensure accuracy. Fortunately, such methods do exist, and surveys remain useful.

Indeed, surveys of parents and children even suggest that some research about traditional television viewing is still relevant and can complement emerging studies on the impact of new media. For instance, one well-designed survey finds that, despite the increase in use of portable computers and interactive apps, much of the mobile-phone content kids consume is in the form of entertainment video from YouTube and Netflix.[4]

Finally, a problem that bedevils old- and new-media research alike is that even the most compelling correlations between behaviors and outcomes do not prove causation. We must therefore be careful not to interpret a relationship between two variables as evidence that one variable caused the other. For example, as we will see, children diagnosed with ADHD consume a lot of media and are especially given to media multitasking. But does this mean that children with ADHD gravitate toward excessive media consumption and media multitasking, or does excessive media consumption and media multitasking exacerbate or cause symptoms of ADHD? As we will see, the preponderance of evidence suggests there may indeed be a causal link.[5] But that link cannot just be presumed. Proving it requires clever research design and careful analysis of results.

Thankfully, we have plenty of rigorous, older research to draw on. Judiciously interpreted, it can and should inform today's raging debate about the impact of screen time and mobile media use on children and adolescents. In particular, when paired with recent progress in our understanding of the implications of new media and related technologies on brain development, traditional-media studies provide insights on what we should expect for children today as they grow into the adults of tomorrow.[6]

iPad Infants and Tech Toddlers (Ages 0–3)

Julie Aigner-Clark was a stay-at-home mother and former high school English teacher. When her first daughter was about a year old, Aigner-Clark started to wonder whether there was a way to expose her to the arts and sciences. She looked around but found the marketplace for infant education lacking. So she decided to make her own curriculum. She borrowed a video camera and, with a few puppets plus the help of her husband, she shot her first film in her basement in 1996. It was the birth of a new concept in media as well as a company that would soon be famous across the United States: Baby Einstein. "Everything

I did in the first videos was based on my experience as a mom," she said in an interview. "I didn't do any research. I knew my baby. I knew what she liked to look at. I assumed that what my baby liked to look at, most other babies would, too."[7]

The idea that you could expose children as young as six months of age to high culture and foreign languages through videos resonated with millions of parents. The first series, also called *Baby Einstein,* was followed by others including *Baby Mozart, Baby Galileo,* and *Baby Shakespeare.* The videos typically were simple montages of puppets, toys, and shapes set to music and poetry read aloud. Parents praised the videos and bragged that their children would sit quietly for hours, watching intently and apparently learning. Baby Einstein products won several awards, including a Best Video prize from *Parenting* magazine. At its peak, Baby Einstein's videos were in nearly one-third of American households with children under two years of age.[8]

Success sparked competition. The UK program *Teletubbies* followed in 1998, as did others with fun and hopefully educational content. After a few years, Aigner-Clark herself wasn't having much fun, facing the pressures of leading a growing business. She sold the company to Disney in 2001 for an undisclosed amount. "The acquisition of Baby Einstein provides Disney with another high-quality brand franchise which serves one of our core customer segments— families with small children," Disney president Bob Iger explained.[9]

A few years later, however, things took a turn. A 2007 paper in the journal *Pediatrics* became the first in a series of research articles that questioned the educational impact of the infant videos. The researchers called more than a thousand parents of children age two or younger to survey them about their child's viewing habits and administer a short version of a well-validated language test used to assess children's early verbal skills. The results were not encouraging for the Baby Einstein franchise or for video exposure in infancy in general. The data showed that each hour of video content infants (age 8–18 months) watched per day was associated with a significant *decrease* in language acquisition. The impact was particularly strong in the youngest children. Not only were infants not learning language at a faster rate, but the evidence suggested that they were falling behind.[10]

In response to the research, *Time* magazine published a provocative and critical article entitled, "Baby Einsteins: Not So Smart After All." Child-advocacy groups and the US Federal Trade Commission also questioned the validity of claims associated with the products. Under pressure, Disney dropped the word "educational" from its marketing. The company followed with an unprecedented offer to refund parents for some of their DVD purchases, a move that was interpreted as a tacit admission that the videos "were not producing Einsteins after all."[11]

Researchers today point to four main reasons to be concerned about any television and video consumption by children prior to age three. First, recall that brain development is most pronounced in the first years after birth, so the interaction among genetics, neuronal growth, and environment is especially critical at this time. There is evidence that exposure to the bright lights and fast pacing of many child-oriented videos can lead to significant effects on attention and cognition. As I discuss later in this chapter, even background television alters the way very young children play with toys and how parents interact with their children.[12]

Second, infants and toddlers cannot actually learn much from videos. Even if young children can grasp what is happening on a screen, they lack the cognitive skills to translate that information into knowledge that is meaningful in the three-dimensional world of reality. This so-called video-transfer deficit likely explains the educational failure of *Baby Einstein* and similar early-childhood audiovisual products. This deficit is probably related to the immature brain in general and the infantile prefrontal cortex in particular, an immaturity that results in poor attentional control, underdeveloped memory capabilities, and an inability to carry out abstract and symbolic thinking.[13]

One of the key pieces of evidence for the video-transfer deficit is a 2010 randomized controlled study of vocabulary learning. Researchers recruited ninety-six families with children between twelve and eighteen months of age and assigned each child to one of four groups. One group watched a best-selling infant-learning DVD at least five times a week without a parent present. Another group watched the same video just as often but with a parent. The third group of kids was tasked with learning the same words only through interaction with their parents—no DVD. A fourth group served as a control and was not exposed to the words at all. The study lasted a month, and the children were tested on their knowledge of the words before and after the study period.[14]

The results clearly show that the toddlers exposed to the video learned no more words than did the control group, regardless of whether a parent was involved. The toddlers who were taught by their parents without the DVD showed the most learning. Interestingly, parents who joined in watching the

DVD and reported that their kids enjoyed the video tended to overestimate how much their children had learned, which, again, was very little. The study makes plain that, when it comes to infant learning, videos are no substitute for face-to-face interaction. Adults who believe otherwise are assuming brain resources that don't yet exist. Social interaction with parents did not effectively supplement video-based learning because, among toddlers, there just is no such thing as video-based learning.

A third concern about viewing habits among the very young comes under the heading of the displacement hypothesis: time spent watching video potentially displaces other more age-appropriate activities such as face-to-face interactions, creative or open playtime, physical movement, outdoor play, and reading, all of which are known to foster brain health in kids. As noted, surveys suggest that some young children today are spending upward of three hours per day with media, an enormous proportion of their waking time spent not doing developmentally important tasks.

Finally, early viewing habits stick with us, fostering potentially serious effects on social, emotional, and physical health across the lifespan. Early exposure wires young brains to cope with emotions and moods using media instead of through relationships and at the expense of much needed lessons in self-control. Habits are built over time and are influenced by emotionally rewarding stimuli that become harder to change as they become more firmly wired in the brain. As researchers Dimitri Christakis and Frederick Zimmerman put it, referring to just one kind of media, "The influence that television might play in this neuromaturational process should not be underestimated."[15]

This influence can result in multiple negative effects that manifest as kids grow up: reduced executive function, mental health issues, higher rates of nearsightedness, and increased levels of obesity, which contributes to other physical health problems down the road. Meanwhile research continues to show that most media exposure for children under three years of age produces no health benefits whatsoever and few if any educational ones; there is simply no getting around the video-transfer deficit. These findings hold regardless of socioeconomic status. So unequivocal are the research results that the American Academy of Pediatrics (AAP) strongly discourages *any* media exposure for children under two years of age.[16]

Yet, more and more kids are spending more and more time watching videos before their first birthday, enabled by the extraordinary ease of access and high-speed mobile digital technologies. Let's focus on one of the most significant harms of excessive early media exposure: a harm that spirals outward to touch every area of social, emotional, and intellectual learning, not to mention creativity and physical well-being: attention deficit disorder.

ADHD and Media Exposure

Dimitri Christakis is a pediatrician and director of the Center for Child Health, Behavior and Development at Seattle Children's Hospital. He is also a major contributor to the AAP guidelines and has been a leading figure in research concerning the impact of media on children. His work includes one of the few studies that measures how television exposure in early childhood affects kids' attentiveness as they age.

Christakis and his colleagues analyzed data from the National Longitudinal Survey of Youth, a study by the US Department of Labor that collects information about individuals over the course of years. This enables researchers to better understand correlations between childhood experiences and later outcomes. Using data from the 1990s, Christakis's team tried to find out if kids' viewing habits at age one and age three predicted attentional problems at age seven. The study found both that kids watched a lot of television—the younger kids in the sample watched an average of 2.2 hours per day, the older kids an average of 3.6 hours—and that more TV time was associated with higher rates of attentional problems. As Christakis explains, "The study revealed that each hour of television watched per day at ages one through three increases the risk of attention problems by almost 10 percent at age seven."[17]

The results were controversial at the time, raising concerns that the authors were minimizing genetic contributions and were blaming parents for causing ADHD in their children. As we consider the results of this study today, we also need to keep in mind that the data are several decades old and were captured before the advent of the internet. While this does not invalidate the findings, we should be cautious about extrapolating to the modern media landscape. The other caveats mentioned above also apply: data were based on parents' reporting about their kids, which is sometimes biased; and this is a correlational study: it does not establish a causal relationship one way or the other between television-viewing and attentional problems. It is possible that children with poor attention spans at an early age are more attracted to and demanding of television exposure. And because children with ADHD can sit quietly and focus on TV, media often gives overwhelmed parents a much-needed break, increasing the amount of media exposure experienced by at-risk children.[18]

These concerns are valid, but they lose much of their force in light of the fact that multiple subsequent studies have now produced similar findings. Reviews of more than fifty scientific studies over the last four decades show a consistent modest but significant effect of media intake on symptoms of ADHD.[19] The correlation appears to have grown stronger in the digital age as well. In 2011, while smartphones were rapidly gaining penetration in the United States and overall media consumption was on the rise, rates of ADHD in children were also on the rise. According to the Centers for Disease Control and Prevention, the proportion of children who have an ADHD diagnosis rose from nearly 8 percent in 2003 to 11 percent in 2011. This translates into a staggering average increase in ADHD cases of 5 percent per year—in line with Christakis's finding that as media exposure increase so does the incidence of ADHD symptoms. And the increase did not stop with young children. A separate analysis during the same period showed that US ADHD rates increased 33 percent for children age 5–9 years, 47 percent for children age 10–14 years, and a shocking 52 percent for teens age 15–17 years.[20]

The data suggest that some children may be more vulnerable than others at an early age, so that some kids can watch lots of TV without ill effects, while many others are more likely to struggle with the consequences. It may be that there is a complex reciprocal relationship between ADHD and media use—the child with risk factors for ADHD seeks out high amounts of stimulating media content, which then confers an increased risk for developing symptoms. Clearly more research is needed to better understand the role of media in early childhood and the longer-term effects of exposure. But what we do know gives us enough to conclude that, when it comes to media exposure, less is more in the early years of brain development.

The Impact of Background Media

Background television is a common feature in all of our lives. Large screens show sports, news, and other programming in gyms, bars, and many places besides. In some homes, too, background television is a constant. Unfortunately, it can have concerning impact on the lives of very young children.

Older siblings and adult caregivers watch plenty of television and stream video while infants and toddlers are nearby, or unintentionally leave the television on in a room where little kids are playing or sleeping. A 2012 survey showed that caregivers were exposing children eight months to two years of age to over five hours of background television per day on average. A 2017 survey found that, in 42 percent of households, the TV is "always" on or on "most of the time."[21]

The impact of background television on young children's experience can be substantial, even if they don't pay much attention to the images and sounds. In 2008 researchers studied fifty children ages twelve, eighteen, and thirty-six months while they played with a variety of toys for one hour—half the time with the television on in the background. The researchers found that the length and quality of playtime were considerably reduced when the TV was on compared to when it was off. It did not matter that the youngest children tended to look at the TV for only a few seconds at a time and typically less than once per minute; all age groups experienced significant effects. Subsequent research also found that the presence of background television reduces the length and quality of parent-child interactions.[22]

Background TV disrupts sustained and focused attention in children (and adults) by eliciting what psychologists call orienting responses. An orienting response follows a visual or auditory cue, a stimulus that suggests novel or potentially relevant information is forthcoming. When the cue arrives, the person experiences a spike of arousal—a signal generated by the brain that prepares us to focus our attention or to act in response to the cue. The authors of the 2008 study suggest that the TV's audio signals and moving pictures, bright light, and large size compete for children's attention while simultaneously arousing them.

Now imagine a situation where a young child spends two to three hours per day using media and is also exposed, for up to five hours per day, to background television. How much high-quality interaction from a caring adult is that child getting? How much less time is available for age-appropriate play, reading books, and face-to-face social interaction—activities known to foster healthy brain development. Meanwhile, thanks to the video-transfer deficit, children three years old and younger gain very little from their viewing habits and, in some cases, fall behind in learning skills required for healthy development.

Of note, the AAP does call out one important exception to its recommended ban on screen time for very young kids: video-chat. With parental guidance, infants and toddlers can benefit from video-chat using applications like Face-time, Skype, Zoom, and Meet, which foster opportunities for social connection, often to loved ones or caretakers in faraway places. In this case, the use of a mobile digital device might be worth the time spent.[23]

The rationale is simple. Video-chat provides something akin to live face-to-face interactions, which is not true of most other media experiences. Video-chat is the exception that proves the no-screens rule for very young children. Children under three years of age learn and develop best through reciprocal face-to-face interactions with attentive, loving parents, older siblings, and other adult relatives and caregivers. The very young need to imitate those around them, repeat what older kids and adults do with their bodies, faces, and speech. Video-chat can provide opportunities to work on these nuanced skills. It also enables reciprocal communication, although it provides no opportunities for comforting touch, which is also valuable for developing neuronal connections in the young prefrontal cortex. Video-chat is not a perfect substitute for in-person communications, but because we are wired to connect, a certain amount of this particular screen time is acceptable, especially when grandparents live far away and parents travel for work. Even kids can exercise their social-bonding skills with video-chat, helping to prepare them for high-quality relationships later in life.[24]

The New Wonder Years (Ages 3–5)

It was a late winter weekend, and I was home alone in Boston with my oldest son, who had just turned three. I had a lot of chores to do and needed a few quiet moments, so I gave him my iPad for an hour. That's the limit, according to the AAP: a child his age is beginning to overcome the video-transfer deficit but should be exposed to no more than an hour of media each day.

My son diligently and enthusiastically watched videos on the YouTube Kids app, which restricts content to help ensure it is age appropriate. For an hour, all went well. But then it came time for him to stop. When I tried to take away the iPad and his videos, I was met with loud screams, tears, and soon after a full-blown emotional meltdown—easily the worst temper tantrum I had ever seen from him.

Initially I assumed he was just tired or having a bad day or was hungry, but after a few more sessions with the iPad and YouTube videos, all of one hour or less, the tantrums continued. My wife noticed the same pattern. This wasn't our son's usual behavior. He would be calm and alert after we read children's books to him for an hour. When we shut off the TV, on which we allowed him to watch long-form professionally produced educational content, he would protest a bit, but he never threw tantrums. Turning off short-form amateur YouTube videos, however, led to total emotional dysregulation.

What we were witnessing was the collision of early brain development and new-media content and technology. Brain development occurs from birth, but it is around age three when these changes really bear fruit. As the brain matures, early language develops and kids begin to become aware of their own identities. The child explores the external world with more energy, intention, and curiosity. Driving these changes are three emerging components of executive function managed by the prefrontal cortex: 1) enhanced working memory, 2) enhanced ability to focus on and shift between tasks, and 3) the ability to inhibit responses to distracting stimuli. The consequence of this developmental leap is that preschool children suddenly can organize their thinking, communications, and behaviors with much more flexibility and control than they could previously.[25]

Now add to the mix the allure of smartphones, apps, and online video content that is curated with an eye toward absorbing kids' attention and mollifying parents—after all, media producers know that parents and caregivers worry about their kids' screen time and wrestle with their own relationships with technology. So content creators and distributors do what they can to give adults a sense of control over the videos their kids see and may even offer features like ad-free videos. (This doesn't mean that kids are protected from marketing, though; an ad-free video of Ryan unboxing toys is still kid-focused marketing.) The result is a particularly challenging and confusing couple of years between the ages of three and five, when parents receive constant invitations to park their kids in front of screens, even as they are just on the edge of developmental readiness. This makes media management in the preschool years particularly difficult and the need for digital literacy of critical importance.

One reason this age is so challenging is that it is the point when on-screen content becomes potentially useful to kids. By 3–5 years old, the video-transfer deficit seen in younger children is surmountable, and some television exposure can complement face-to-face learning. A large body of research has found that high-quality, long-form, professionally produced educational content such as *Sesame Street* and *Dora the Explorer* can help children of appropriate age obtain language and literacy skills. But what about new media content online and so-called educational software?[26]

In theory, nothing prevents new media from delivering old-fashioned linear content that is carefully designed with kids' development and education in mind. And nothing prevents software developers from creating sophisticated and programmable interactivity that supports learning. Indeed, the promise of educational software is considerable, as interactive programs can move along according to children's readiness, structuring activities and feedback to offer appropriate challenges that reinforce and reward learning. The availability of highly customized learning on ubiquitous portable devices could be good for kids and parents, facilitating knowledge acquisition anywhere and freeing busy caregivers to look after their own work, needs, and adult development, secure in the understanding that their young ones are actually nurturing their brains.

Unfortunately, the world of developmentally appropriate new media is not the one we live in. The promise of interactive software is real but unrealized: appropriately designed e-learning software has been shown to facilitate effective learning by preschool children, yet there is little evidence that commercially available products embrace these design principles. A 2015 review of hundreds of apps labeled "educational" for toddlers and preschoolers was discouraging at best.[27]

The review found little reason to trust that any of the apps worked well. Few were based on any claimed theoretical model or formal testing, meaning that no one had actually tested whether the apps taught young children anything of use. What testing was acknowledged was overwhelmingly usability testing, focused on whether the software worked or delivered "appeal" or "likability" that is common in marketing to adults. Noticeably absent were tests of learning efficiency tied to research outcomes consistent with children's needs. And apps did not facilitate activities that are known to be essential to childhood learning and development. For example, less than 10 percent of apps encouraged users to share their experience with a parent or caregiver, even though there is ample evidence that active parental involvement in both digital play and book reading improves children's learning.[28]

It is no wonder that supposed expert ratings of the apps varied greatly, so that one list of recommended apps would look nothing like another. This ostensibly educational software was designed to be liked, not to teach. It was created to subjectively appeal to and engage young users, not to achieve more objective learning goals. In such a situation, these apps appear to help parents but in fact only add to the confusion of raising kids in the digital age.

The overarching goal of businesses that create interactive digital experiences, in particular gaming and online video, is behavioral reinforcement that keeps users coming back. Platforms thrive on having the greatest number of users and the longest duration of use. Software developers therefore focus on creating compulsion loops using endless autoplay features and infinite content recommendations rather than on learning goals and literacy. Further complicating matters, animations, musical interludes, and other bells and whistles make an experience appealing but serve mainly to titillate and distract users; rarely do they facilitate skill acquisition. Actual learning requires pacing, proper timing of rewards, and clear learning objectives, but children don't need any of that in order to engage with "educational" content. They will watch and play for the sake of stimulation and entertainment.

Now let's return to my son's emotional meltdown. I'm hardly unique in turning to an iPad to babysit my kids every now and then. Studies show that parents and caregivers often rely on mobile devices when they need a break from childcare. These same parents and caregivers may be hesitant about screen time and may express concern or guilt about putting a device in their kids' hands. But sometimes it feels like there actually is no choice. Whatever guilt I felt, I gave my son that iPad, just as so many adults do.[29]

Along with many parents and caregivers, I have discovered that the structure of new media makes it particularly hard to extricate kids from the digital playpen. Reengaging kids is often hard because the digital tasks on mobile devices are designed to engage without the natural breaks and end points built into traditional, professionally produced television programming. Autoplay, endless recommendations, skip functions, and video games enabling tens of hours—or even limitless—play ensure that new media can only be interrupted, creating greater tension when it is time to stop. Research confirms what caregivers have intuited: new-media screen time leads to especially unmanageable tantrums among toddlers and preschoolers.[30]

Some academic and industry leaders have recommended changes that would address these concerns and enable new media to fulfill its promise of helping kids realize clear learning objectives. For one thing, if an app has bells and whistles, these should reinforce the substantiated learning objectives, not distract or merely stimulate the user. For another, apps should encourage caregivers to engage alongside their kids, to enhance knowledge acquisition and extend interactions more naturally into the nondigital environment. At a minimum, software should provide parents reports on the activities their kids are working on, so that adults can monitor their children's progress and learning. And the app or online experience should incorporate stopping points, set to kick in by default, to signal when it is time to end. This would encourage children and caregivers to pause the experience and return to the real world. However, very few digital experiences adhere to these recommendations.[31]

My wife and I learned a valuable lesson from that weekend tantrum and through subsequent trials and tribulations with media in our children's lives. Now we seldom use the iPad and online short-form video, and only as a reward for good behavior. When one of our children uses the device, we set clear expectations about how long the experience will last and give ten-, five-, and two-minute warnings so that our kids know the experience is coming to an end. Mostly we let them consume their one-hour-per-day of media with one of us present, on a large screen or television in a common room, or we steer them toward long-form, professionally produced educational content, even though usually they each want to watch something different. When we travel, we use portable DVD players without access to the internet. This limits their choices, gives us control over their viewing, and helps ensure that they don't become sucked into the bottomless pit of online recommendations. And we only allow them to use apps that serve a clear, age-appropriate learning objective; incorporate stopping points; encourage parental interaction; and limit bells and whistles.

Learning to Read in the Digital Age

One consequence of the rapid neural growth of very young children is that they are like sponges when it comes to learning, soaking up practically everything in their environment. But not all learning is equal. As the child enters the preschool years, the ability to learn accelerates amid explosive brain growth. The child begins to explore, socialize, and develop a theory of mind critical to forming quality relationships. This is often when parents and other caregivers introduce more structured learning, with a particular focus on learning to read.

Reading is a skill possessed uniquely by humans. Estimated to have evolved approximately 5,000 years ago, it is a recent advancement that did not exist for most of our species' time on Earth. But its impact on our technological development, social structures, and domination of the planet cannot be underestimated.

Because reading is a skill invented late in our species' evolutionary history, none of the neural structures hardwired into our brains are genetically programmed to enable us to read. Instead, brain networks developed for other functions are recruited and repurposed for the task of interpreting letters into words, words into sentences, and sentences into meaning. This is why reading competency comes only with purposeful effort invested over time. Learning to read requires intentional practice and lots of repetition. And because reading rewires our brains, it is important to practice when our brains are at their most malleable: during the period between infancy and age five. Evidence shows that this developmental period, critical for acquiring so many skills, is also essential for competent lifelong reading.[32]

That's why I try to sit and read with my children as often as possible. Shared reading with adults is a potent means to enhance the reciprocal verbal and visual stimulation necessary to develop the neuronal networks required for learning to read. And research suggests that the lap sitting and other nurturing behaviors associated with reading picture books to very young children not only enhances reading skills but also the child's attachment to the adult. Studies also show that kids exposed to more reading in the preschool years are more motivated to read in the future, which is associated with increased academic achievement and better health. By contrast, other data suggest that reduced early reading is correlated with reduced readiness for kindergarten, decreased academic performance, and even increased rates of obesity, which leads to other poor physical health outcomes. Not surprisingly, the same AAP that recommends against nearly all screen time before age two also strongly supports introducing reading as early after birth as possible.[33]

This is not a simple task, though. As adult readers we often take for granted the ability to translate letters on a page into meaning. Reading involves several complex mental processes, which tax a child's limited cognitive resources. Readers need to keep some amount of information in their minds as they examine letters and words using visual working memory. They must be able to ignore extraneous stimuli and maintain focus using attention and inhibitory control. And they must derive meaning from abstract symbols and keep track of meaning over time to establish reading comprehension and fluency. All of these skills are often lumped under executive function and are largely managed by the prefrontal cortex. Accordingly, impairments in reading—whether from lack of exposure, learning disabilities, or ADHD—look like altered activation, lost connectivity, or some other failure within this critical area of the brain.[34]

As much of a challenge as learning to read presents, that challenge becomes more daunting in the digital age. Young people devote less time to reading as rates of digital screen use rise. But the problem is not just that time spent on screens isn't time spent reading—or doing other developmentally valuable activities like physical exercise, drawing, playing with other children, or socializing with adults. There is also evidence that large amounts of screen time at young ages harms the brain in ways that make reading more difficult to learn and less rewarding.[35]

As discussed earlier, increased screen time contributes in modest but meaningful ways to attentional problems in school-age children, perhaps due to a lag in maturation of the prefrontal cortex caused by high exposure levels. Kids who have a hard time paying attention also have a hard time focusing on reading and therefore struggle to develop fluency and comprehension. Their attempts to read are more likely to lead to frustration, as well as the sense that reading, because of reduced comprehension, is boring and otherwise not worth doing. This results in a negative feedback loop in which little is gained by reading, which leads to hard feelings, which leads to fewer attempts to read, which leads to frustration and fewer gains from reading, and so on.

Recent neuroimaging evidence suggests there is a very big difference between the brains of children who read and are read to compared with the brains of children who spend less time reading and consume large amounts of media. More generally, we know from brain-imaging studies that children who experience very high levels of screen time also experience cortical atrophy—reduced gray matter and white matter in the cerebral cortex, reduced neuronal connectivity, and decreased brain thickness in language areas.[36] Here I focus on the evidence of screen time–associated brain changes as they relate to reading acquisition.

Some of the key evidence comes from John Hutton, a pediatrician and neuroscientist at Cincinnati Children's Hospital. Hutton and his colleagues carried out two studies that together suggest that screen time can be disruptive to the development of reading skills. In the first study, Hutton and his colleagues used fMRI to scan the brains of four-year-old girls while they listened to recordings of age-appropriate stories being read to them. This was followed by an uncoached mother-daughter session, in which the girls' mothers could choose to read to them a different age-appropriate story from a picture book. The reading

sessions were videotaped and coded using standard metrics for levels of interactivity, intimacy, and engagement as well as the type and quality of dialogue
between the mother and daughter.

Researchers found that maternal reading scores—the overall quality of the
interactive, face-to-face sessions—correlated with higher levels of activation
in brain areas associated with reading and language while the child was listening to stories in the fMRI scanner. This suggests that high-quality adult-
child interactions during reading help to promote what Hutton calls "reading
readiness"—these interactions train the brain for literacy. Given what we know
about reading, the areas of brain activation were not surprising; they included
several areas of the prefrontal cortex. And given what we know about the role
of the prefrontal cortex in emotional regulation, it begins to make sense that
kids tend to be calm and relaxed, but still engaged, when adults read to them.
Parental reading to children has benefits on an emotional level, especially when
the parent takes care to ask questions, explain material, and prompt the child
to think about what's on the page. Hutton and his coauthors conclude, "It is
reasonable to speculate that children with greater experience reciprocally engaging in shared reading may be better equipped to form stronger social-
emotional connections between stories and their own life, and with the caregivers who read to them."[37]

So there is reason to believe that high-quality shared reading with adult
caregivers trains kids for reading comprehension and emotional well-being.
What about the impact of modern media devices? The second study from
Hutton and his colleagues used brain scans of children 8–12 years old, an age
when people are expected to be fluent readers. These kids' parents separately
told the researchers how many hours their child spent on reading books, newspapers, and any other material for pleasure and how much time they spent
using screen-based media (including smartphones, tablets, laptops, desktop
computers, and television). The researchers focused on the same areas of the
brain in this study as they had in the previous, but this time looked to see
whether kids who spent more time reading than using screen-based media
showed stronger connections in a reading-associated neural network that includes the prefrontal cortex, vision- and language-processing centers, and certain other brain regions.[38]

More time spent with screen-based media was associated with less functional
connectivity in the reading-associated network. In contrast, more time spent

with traditional reading was associated with more connectivity in the same network. In light of the findings of both studies, Hutton and his team hypothesize that learning to read requires more active information processing than does most screen-based media consumption, particularly in early years of development. In other words, reading demands more of the prefrontal cortex and more connectivity with the rest of the reading-associated network. Readers need to orchestrate attention, information processing, visualization, and imagination. The emergent theater of the mind unique to reading books reinforces these critical brain connections. Using screen-based media does not.

The results from these two studies, carried out in 2017 and 2018, offer some of the first direct evidence that the brains of young children who spend more time with media primarily through screens suffer compared to the brains of those who spend more time on independent reading. The findings enhance our understanding of reading's positive contributions to neuronal development, emotion regulation, and secure attachment. And if indeed screen time—even interactive screen time—requires less effort by fewer neural networks, we can also expect that several critical brain areas, including the prefrontal cortex, will show weaker connections in maturing kids who turn more to screens than to books.

But what about e-readers? Digital books are a bit of a wild card. Some studies show that preschool children learn content and vocabulary equally from digital and print books, as long as formal features of the digital books support learning rather than distract from it. But there is also evidence that digital books undermine the benefits of shared reading with adult caregivers.[39]

There are two overarching reasons for this. One is that digital books tend to engage children in more solo reading, so that shared reading experiences simply happen less often. The second reason is that the quality of parental interaction is reduced. For example, when reading digital books with young children, parents carry out fewer reciprocal reading strategies, such as labeling objects in the book, asking open-ended questions about the characters, and commenting on the story beyond what is contained in the pictures and words. Instead, when using electronic books, parents more frequently comment on the digital device itself, with more utterances such as "push this" and "tap that." When using e-readers, then, kids are at risk of losing out on the active parent and caregiver involvement that is critically important for learning and development of the prefrontal cortex. Christakis is fond of saying that young children "need laps more than apps." The research appears to support this sentiment.[40]

The Rise of Media Multitasking (Ages 5–12)

Anxiety increases for parents and children in the school years. As children's brains continue to mature during the early grades, new environments, schedules, friends, and learning demands emerge. Children also begin to invest more time in their technology habits, creating challenges for the developing prefrontal cortex. One of these challenging habits is media multitasking.

As we have seen, media multitasking involves the simultaneous use of multiple devices, such as watching a TV show while using a smartphone to access social media. One component of rising media multitasking is simply increased use of media generally. A 2010 report from the Kaiser Family Foundation found that US children ages 5–8 average about three hours of media exposure per day via TV and online videos, video-gaming, listening to music, browsing the internet, and using social media. By ages 8–12, that number nearly doubles. And media use is skewed toward kids whose families are less well off. The report found that, among the 5–8 year-old group, parents with higher income or education level reported that their kids spent significantly less time consuming media. There may be many reasons for this difference. Parents earning less may need to work multiple jobs, giving them less time to focus on their kids and leaving their kids to babysit themselves with media. Less well-off parents may also be unable to afford high-quality childcare—another reason their kids use more media. An additional factor is probably lower awareness of the AAP guidelines and reduced digital literacy compared with better-educated and higher-income parents.[41]

Other AAP recommendations that are consistently ignored include guidelines on media use before sleep and background television. As discussed, more than 40 percent of parents report leaving the TV on in the background "always" or "most of the time" in their homes. And in 2017 nearly half of parents reported that their kids use media technology in the hour before sleep.[42]

These findings suggest that rapid changes in media access and the proliferation of digital devices are outstripping parents' ability to keep up with their children and preventing parents from applying some basic limits to protect their kids' developing brains. In 2011, 52 percent of US children had access to a mobile media device. By 2017 that number had practically doubled, with nearly all (98 percent) of children age eight and under living in a home with a mobile media device. With all those devices around, media use increased commensurately. "We thought children's screen time couldn't rise any higher," says Don Roberts, a Stanford University communications professor involved in the research. "But it just keeps going up and up."[43]

Alongside shifts in access to mobile media technology have come shifts in usage patterns. Combine the use of computers for schoolwork with the ubiquity of mobile media devices, increased interest in video content, and more gaming and social media options than ever before, and we have seen an incredible rise in media multitasking.

In the 5–9 year-old age group in the United States, nearly one-quarter of media time is spent using more than one screen simultaneously. In the 8–10 age group, average media use per day is nearly five and a half hours, but total media exposure is nearly eight hours, suggesting very high amounts of media multitasking must be occurring. This typically involves playing a game on a smartphone with the TV on or doing homework on a computer while streaming music, checking social media, or otherwise using the internet. The phenomenon is not isolated to the United States, as studies show high amounts of childhood media multitasking in countries from Russia to Kuwait.[44]

The early school years demand special attention because this is a period of significant development. Important elements of our physical and mental lives become ingrained—not immutable, but harder to change—as durable personality traits form. At this time, cognitive, emotional, and social skills are still developing rapidly, albeit in new ways. Social attachment is shifting from caregivers and family members to friends and teachers. In addition, habits surrounding diet, physical activity, and sleep begin to consolidate. These changes have long-term implications for healthy brains and bodies.[45]

Given the importance of these changes, it is not surprising that academic and industry researchers are increasingly paying attention to the impact of rising rates of media multitasking on school-age kids. This sort of research follows what is by now a familiar protocol. First investigators ask children and their parents to fill out questionnaires probing the children's media habits. Then the kids are divided into groups of heavy media multitaskers and light media multitaskers and asked to do standard tasks in controlled environments that allow researchers to test their abilities and skills in one or more domains. Finally, the researchers look for differences between the two groups related to media use. The findings are concerning. Again, the direction of causality is a challenge. Do children who indulge in heavy media multitasking develop certain cognitive or emotional characteristics that affect their skills, or do cognitive and emotional characteristics affecting skills also leave children more attracted to media multitasking?

As in other cases, we can overcome some causality problems by doing lots of studies on a theme using a range of methods and analyzing the results. What we find from such reviews paints a disconcerting picture when it comes to media multitasking and skill acquisition among school-age kids. More research is needed, but a 2017 review of the literature concludes, "the weight of evidence overall points to [heavy media multitaskers] demonstrating reduced performance in a number of cognitive domains relative to [light media multitaskers]."[46]

The review points to a range of mental abilities and psychological functioning that suffer as rates of media multitasking increase. Heavy media multitaskers between the ages of five and twelve years experience deficits associated with ADHD, including poor short- and long-term memory. They also experience what is known as filter failure—the reduced ability to filter out extraneous information. Relatedly, they face difficulty keeping on a task for long periods of time, revealing a deficit in focused attention. Getting any of these cognitive skills right requires a healthy prefrontal cortex. In addition to cognitive deficits, the reviewed studies show that heavy media multitasking is correlated with higher levels of impulsivity, more sensation-seeking behavior, and higher rates of social anxiety and depression.

The consequences of reduced cognitive abilities and increased emotional struggles during the school years are potentially serious. Of course, it is always worth remembering that childhood problems are often correctable and indeed self-correcting in some cases. But we also know from our own lives that even modest levels of distraction and stress can affect our ability to learn and work.

Evidence backs up experience. Another review examined the impact of media exposure, including media multitasking, on the academic performance of young students, adolescents, and university students, with grim conclusions. The review examined performance measures such as course grades, grade point averages, and test scores as well as measures of study habits and attitudes toward school-based learning. Of the studies that looked at media multitasking in the classroom or while studying, 70 percent found a negative impact on academic performance. The size of the impact varied, with some multitasking activities showing a larger influence than others. For example, the simultaneous use of social media platforms while doing schoolwork, such as text messaging and using Facebook while doing homework assignments, was particularly detrimental. Across the findings, lower grades were associated with social media use while doing schoolwork; the consistency of this finding suggests that social media is especially disruptive to learning. Perhaps the lure of online interaction, and therefore the time spent on it, is especially great because of the deftness with which social media designers hack our reward systems. Or maybe the interactive nature of social media—compared to passive television viewing or music listening—is particularly disruptive of cognitive processes.

Alongside studies of media multitasking's effects on schoolwork, the review analyzed effects on performance in controlled laboratory environments. In studies like these, experimenters typically assign participants a task to complete, with or without a distracting media device on hand. The review examines multiple studies in which school-age participants were asked to read a standard, age-appropriate text and were then scored on their ability to comprehend the story and recall details. These studies consistently show that participants' comprehension and recollection of details suffer in the presence of distracting media devices.[47]

The results strongly suggest that media use during academic activities significantly hinders students' learning by diminishing executive function and cognitive control managed by the maturing prefrontal cortex. What is the long-term impact of reduced executive function in childhood? While more research is needed, one landmark study published in 2011 on the impact of self-control (a broad measure that includes executive function) suggests serious effects. The researchers found that children with low levels of self-control at ages three and five went on to have negative outcomes in three domains as adults: health, wealth, and criminality.[48]

The researchers looked at a group of more than a thousand children born in a single year, in a single city in New Zealand, and followed them until adulthood (age thirty-two). With respect to physical health, the children who had poor self-control went on to have higher rates of obesity and increased risk of periodontal, cardiac, and respiratory disease. As adults they were also at higher risk for alcohol and drug problems, and they were more likely to conceive a child raised in a single-parent household. Adults who had poor self-control as kids were also more likely to be in trouble financially, with lower savings rates, higher amounts of credit card debt, and fewer financial assets such as real estate, investment funds, and retirement plans. Finally, children who scored lower on self-control were also more likely to be convicted as adults of criminal offenses.

The results were unaffected by gender, controlled for the presence of ADHD, and were also independent of childhood IQ and family socioeconomic status.

It is remarkable that a single metric—in this case a measure of self-control, for which a healthy and functional prefrontal cortex is critical—could consistently predict so many outcomes decades later. One commentator on the report concluded, "There may be no such thing as 'too much' self-control." Clearly too little has severe consequences later in life.[49]

In interpreting studies like these, we need to keep in mind that many factors can influence self-control. These include quality of parenting, social connections and attachment levels, adverse events, childhood experience of neglect, and chronic stress. In addition, these research findings concern kids who grew up before the smartphone revolution and the shifts in media behavior that are at the core of this book. But the results remind us of the importance of nurturing a healthy prefrontal cortex to improve future outcomes. The results also foreshadow some of what we will learn about the behavioral and neurological consequences of modern media technology for tween, teen, and adult brains.

IN BRIEF: THE MARCH OF MYOPIA

Alongside attention deficit, loss of emotional control, and other executive-function deficits related to a weakened prefrontal cortex, there are physical consequences to some of our shifting technology behaviors. One of these is poor vision. Our changing relationship with technology is provoking a global epidemic of nearsightedness—myopia, as it's clinically known. It is estimated that nearly 90 percent of Chinese teens and young adults are nearsighted. A 2015 study found that 96 percent of nineteen-year-old South Korean men are diagnosed as nearsighted. In the United States and Europe, nearly half of all young adults have myopia, about twice the prevalence of fifty years ago. Some estimates show that nearly a third of the world's population could be afflicted by the next decade.[50]

Myopia occurs when the eyeball becomes overly elongated, causing light entering the eye to focus in front of the retina, rather than directly on it. This results in blurring of faraway objects, though not objects up close, hence the term "nearsighted." Diagnosis is most common in the late teenage years. The eyeball elongates naturally throughout development, as the eye grows and changes shape. But elongation need not result in poor vision.

The dramatic spike in myopia in recent years led scientists to question long-held assumptions about the causes of nearsightedness. Historically eye strain from excessive close-up reading was thought to contribute to the loss of distance vision. Later theories considered other factors such as genetics. Indeed, twin studies showed similar rates of myopia in twin pairs, and biologists have found some genes associated with the disorder. But follow-up studies suggested that genetics alone could not explain the staggering recent increases in myopia. There had to be another factor at play.[51]

In 2007 a candidate for that other factor emerged. Researchers tracked more than 500 US children with normal vision over a three-year period. They found that, as expected, genetics are a factor, as parental myopia correlated with nearsightedness in children. However, they also found another potential cause: spending more time indoors. Researchers found that kids who spent more time outdoors had lower rates of myopia. It wasn't reading that contributed to myopia; surprisingly, researchers also ruled out screen time as a direct cause. What mattered was just being inside, regardless of what one was doing.[52]

Scientists continue to investigate various mechanisms that might explain why staying indoors contributes to myopia, but there are two likely factors. First, exposure to the outdoors encourages distance viewing and accommodation toward faraway objects, which helps to shape the eye over time. Second, exposure to bright light is known to protect against myopia, and the outdoors are far brighter than our houses, schools, and workplaces. Even on a cloudy day, the eye is typically exposed outdoors to around 10,000 lux, the international unit of illuminance. A sunny day can generate up to 100,000 lux. This compares with an average indoor lighting experience of 500 lux. Several intervention studies strongly suggest that children who spend one to three hours per day outside have significantly reduced risk of developing myopia later in life.

The mechanism appears to be related to retinal dopamine. Retinal dopamine is the same dopamine that works in the brain's reward centers. However, in the eye, dopamine plays a different function: it signals specialized cells in the eye to shift from detecting low light to bright light. This switching happens multiple times per day, but the frequency of switching is reduced with prolonged exposure to dim, artificial indoor lighting. Researchers think that the lack of exposure to bright daytime light disrupts retinal dopamine release, contributing to maladaptation of the developing eye shape.[53]

More research is needed both on the mechanisms of myopia and on the sources of the recent staggering increase in nearsightedness. But it is reasonable to speculate that if time spent indoors contributes to myopia, then the profusion of digital distractions contributing to time indoors is probably harming our vision. It is true that mobility allows us to use screens everywhere, but everywhere doesn't necessarily mean outdoors. And, in general, screen use drives work and play inside. Encouraging children to spend a couple of hours a day outside to protect their vision—and realize other benefits of physical play and time spent in nature—is a reasonable course of action.

IN BRIEF: OBESITY, MEDIA, AND CHILDREN

Childhood obesity is a growing problem worldwide. In the United States, the incidence of childhood obesity has tripled since the 1970s; recent statistics classify one in five children under the age of nineteen as obese. There are many potential reasons for the increase in weight, including greater access to high-calorie foods, changes in daily routines, and economic trends.[54]

One variable with a consistent and potent link to childhood weight gain is high levels of media use. One US study of 10–15 year-old children over a four-year period showed a strong relationship between weight gain and hours of television viewed. TV viewing accounted for 60 percent of the weight gain over time, with other factors, such as socioeconomic status and access to healthy foods, accounting for the remaining 40 percent. Another study looked at the influence of childhood media consumption habits on weight gain in young adulthood. The results showed that a significant amount of weight gained by early adulthood was accounted for by watching large amounts of television as a child, suggesting that some of the impact of childhood media consumption comes years later.[55]

While the relationship between screen time and obesity seems compelling, the issue of causality remains. It could be that children who are more prone to consume excess calories and gain weight are more likely to consume media. One way to establish the direction of causality is to use a randomized controlled trial, in which researchers randomly assign participants to an exposure group or a control group. The exposure group is subject to some intervention, while the control group is not. Then researchers look for differences in outcomes over time.

One randomized controlled trial looked at a seven-month intervention for third and fourth graders in two California schools. One school was randomly selected to have teachers trained in a media-reduction curriculum, which would then be delivered in regular classrooms. The intervention included a television time manager and offered training on media-behavior modification and educational newsletters for parents. The other school received no intervention.

Over the period of the study, the students in the intervention school significantly reduced both their media consumption and the number of calories they consumed in front of a media device. They also had significantly less weight gain. A follow-up assessment indicated that the impact of the intervention persisted for two years.[56]

These are important studies that should pique our curiosity. Why would an intervention designed to reduce media consumption also reduce calorie consumption and weight gain? The initial and obvious assumption pointed to the displacement hypothesis: time spent watching television was sedentary time, displacing physical activities. However, researchers struggled to find evidence supporting this hypothesis, as studies showed that children's screen time did not necessarily replace time spent doing physical activity. So scientists explored alternative theories. Today's leading theory, backed by substantial evidence, is that media induces poor eating habits in children. Kids who consume more media eat a larger amount of lower-quality food: they eat more high-calorie processed snacks, drinks, and fast food and fewer fruits and vegetables. Studies also show that kids who consume more media consume more calories overall.[57]

There are many possible explanations for this finding, but one is that media viewing distracts us from the normal feelings of satiety that stop us from eating. Satiety is triggered by a complex interaction between the digestive system and the brain.[58] The theory is that media consumption distracts our brains to the point that we don't realize we are full, and without the proper neurological signaling, we keep on eating.

There is also evidence from laboratory studies that food advertising can encourage unhealthy eating while watching TV. One study exposed a group of children ages seven to eleven to cartoons that included food advertising, and another group, of the same age, to cartoons that omitted the food ads. The children were all given a bowl of Goldfish crackers to snack on while watching. The children exposed to the food advertising ate 45 percent more crackers than the children not exposed to the food advertising. After controlling for other variables, it appears the excess eating was a direct result of viewing the commercials marketing food.[59]

As advertising to children increases along with increased media consumption across multiple devices, the ability to recognize food advertising becomes more difficult, especially online. For example, advertisers often disguise food product endorsements, including in so-called advergames—games directed at children and designed to encourage them to eat commercial food products.[60] The lesson here is that parents and caregivers have to be mindful not just of their children's overall media consumption but also of the types and quantities of food they are eating, especially as they engage with media, communications, and information technologies.

5

TWEENS, TEENS, AND TECH

Adolescence is a time of incredible creativity and exploration, as teens begin to separate from parents and depend increasingly on peers to define their sense of self. Social pressures grow, and teens experiment with ways to fit in. It can be a period of uncertainty and a certain amount of chaos—hence the stereotypes, not without some merit, of teens hijacked by raging hormones, engaging in high-risk activities, and falling prey to irrational judgments.

This developmental period is also especially mysterious. While the behavioral changes associated with the shift from middle childhood to adolescence were well documented by the turn of the last century, only in this century have we begun to understand the underlying mechanisms of the teen brain. We have learned that the teenage brain is not a defective or half-baked version of the adult brain. It is just entering a final, unique stage prior to full maturity.

This knowledge comes in part from a remarkable series of analyses and a triumph of modern brain mapping. A group over 800 young people ages eight to twenty-two completed a series of brain scans and sophisticated neurocognitive tests that measure executive function. Researchers found that the results on the neurocognitive tests improved steadily with age. This was expected; we know that as the brain matures, executive function improves. What was less expected were the brain changes associated with improved executive function through the teen years. "We were surprised to find that the developmental refinement of structural brain networks involved [both] increased modular segregation and global integration," said Ted Satterthwaite, a professor of psychiatry.[1]

What does this mean? The secret underlying the teenage brain is that it is simultaneously doing two things that might seem incompatible: it is becoming both more modular and more interconnected. As individual regions grow in specialization, their connections with other regions also develop and strengthen. This finding is remarkable because it demonstrates how our understanding of brain function in general is evolving over time. It also dramatically alters how we should interpret the operations of the teenage brain.

In the early days of neuroimaging, researchers weren't looking for networks or connectivity between brain regions. They focused more on areas of the brain that would "light up" in response to some task or stimulus. Magazines and news-papers reported on this exciting new capability, and soon the public and scientists alike were enamored by the latest findings showing which part of the brain was active and, therefore, responsible for some human ability or fallibility. But critics charged that some of these interpretations were overly speculative—a type of modern-day phrenology. They coined the term "reverse inference" to describe the tendency to link functional capabilities to a few hot spots of statistical significance appearing in brain scans, often on the basis of one study with a small sample size.[2]

As research techniques and technologies became more sophisticated, the interpretations became more nuanced. Neuroscientists began to realize that not only was activity in parts of the brain important, but most likely inactivity was, too. And so was neural connectivity across brain regions. One might pause and wonder why modularity and connectivity were presumed to be at odds, but this is actually common in nature. Most organic systems lose connections to each other as they become more functionally segregated and specialized. The brain is different, though. Analyses of neural networks challenged the presumption that the brain is primarily segregated into discrete areas of functional significance.

It turns out, as this adolescent-brain study showed, the human brain is capable of more than one type of developmental change at a time: it can be simultaneously increasingly modular and increasingly networked. The gains in executive function found in the research were associated with increasing regional specialization and with more interconnectivity—certain pathways of integration across modules "thicken." This wondrous and species-specific de-sign helps to optimize human cognitive performance.

This study found a network of connections involving the prefrontal cortex that was strongly associated with executive function regardless of age. This makes clear that while the prefrontal cortex is deeply important to executive function, it can't do the job by itself. Specific connections across regions are also implicated in enhancing the cognitive skills orchestrated by the brain's conductor. These connections become more robust during adolescence, as multiple areas of the brain move into a final stage of maturity.

How does the brain accomplish this amazing organizational rewiring? Neural and cognitive maturation are programmed into our genes, of course. They are also driven by hormonal changes. But while this process is baked into our biology, it is not insulated from environmental forces. Parental and caregiver guidance, along with the influence of teachers, other adults, and peers affect how the map of the brain changes during the teenage years.

There are two basic mechanisms involved in the massive rewiring associated with adolescence. The first is the selective elimination of neurons, known as neuronal or synaptic pruning. Pruning removes extraneous and weakly reinforced brain cells (neurons) and their connections, via synapses, to other brain cells. As one neuroscientist describes it, pruning is akin to sculpting. The reduction in unneeded cells and unused connections helps give rise to many of our most advanced cognitive skills. It is like Michelangelo's *David* emerging from a block of marble. Pruning sharpens our capabilities on our path to full maturity.[3]

At the same time, a second process of accretion and densification is underway. More adaptive and frequently used neuronal connections are strengthened through neurogenesis (the creation of new neurons) and neuroplasticity (the construction of new networks). The outcome is improved coordination of the more specialized and mature brain areas and reduced interference from other less relevant neurons, enabling advanced impulse control and higher executive function. This is an elegant evolutionary solution to the challenge of brain development, balancing genetic and environmental forces to give rise to the sophisticated and nuanced capabilities of the adult brain.[4]

But ask any parent to describe their teenage son or daughter and the word "elegant" won't likely be used. That's in part because a major consequence of this unique maturational process is a significant lag in the development of the prefrontal cortex behind that of two other critical centers in the subcortex—the lower portion of the brain. One of these subcortical areas of runaway growth includes the emotion centers, which give color and meaning to the world and tag information for relevance. The other area, in close proximity to the emotion centers, includes the dopamine-based reward centers that drive many of our motivations and instincts. So the fully blossoming, hormonally charged emotion and reward centers race ahead of the prefrontal cortex and its moderating influence.

Not until a person's mid-twenties does the prefrontal cortex fully mature, catching up to the subcortex and effectively challenging the impulsivity and risk-taking behaviors that arise from the emotion and reward centers. This has a lot to do with why tweens and teenagers can at once be so creative and out of control, and at times irrational. They are essentially wired for a unique period of growth that comes with too much gas and not enough brakes.

This developmental lag is inevitable; the more evolutionarily complex and sophisticated prefrontal cortex takes longer to mature than these older areas of the subcortex. But, again, biology alone is not destiny. Environmental factors, including stress and adverse events, can affect the course of this period of development. As it happens, other important environmental factors include the media we consume and the ways we communicate with one another. In the digital age, these environmental factors are becoming decidedly more influential and consequential to the developing teenage brain.

New Tech, New Teen Brains and Behaviors

When I was a teenager, a television was an entertainment device, a phone was a communication device, and a computer was a productivity device. Today's teens expect every device to be all these and more. They spend more time looking at and interacting with mobile media, communications, and information technology devices than any prior generation in human history. Period.

All that device time has given tweens and teens access to myriad possibilities—not only entertainment, communication, and information but also content creation and commerce. From the perspective of an adult who remembers when phones hung on walls and receivers were tethered with cords, teens nowadays can seem without limit as they explore new online worlds and ways of connecting.

This gives rise to the image of today's teens as high-powered multichannel communicators and multitaskers. They can simultaneously surf the web, exchange instant messages with friends, and play games on a single mobile device. The digital native moves seamlessly across tasks and screens, her life and her peers' lives thoroughly integrated into the cyber-universe. But does this image reflect reality? And what do the changes in the ways teens navigate the digital world mean for their psychology and their brains?

First, some eye-popping statistics. Media consumption jumps significantly in the teen years; a 2019 survey finds that US teens spend, on average, more than seven hours per day using media for purposes other than schoolwork, while preteens (ages 8–12) consume nearly five hours per day (media here include television, online video, computer gaming, social media, browsing and reading online, and listening to music). This extraordinary level of consumption is possible in part because access is ubiquitous. The same survey showed over 50 percent of Americans own a smartphone by age eleven, and another survey by the Pew Research Center showed that 95 percent of US teens have access to a smartphone, whether or not they own one.[5]

Clearly teens use media for many purposes. Some are heavy gamers, others spend more time watching videos, and a few even use digital technology for reading—that last group was the smallest in the 2019 study. What is less variable is the use of digital media for communication. The digital age has altered how all of us communicate, but teens have seen the most radical shift. Teenagers are the heaviest users of text and instant messaging, sending more than a hundred messages per day on average, ten times more than adults age fifty-five and older by some estimates. Outside of school, teens are more likely to communicate by text than by any other means, including face-to-face communication.[6]

Teen digital communications are also highly fragmented, as conversations take place via not only text messaging but also on chat apps such as WhatsApp and Kik and video-sharing platforms like Live.me, YouTube, and TikTok. It is not uncommon for teens to have multiple conversations going with different groups of friends simultaneously, using multiple apps and a range of social media services far beyond those their parents might know. Instagram, Snapchat, and Whisper are just a few of the platforms other than Facebook and Twitter that have grown popular among teens, and more come online seemingly daily. Meanwhile, all this communication occurs while teens are watching videos, streaming music, or doing homework. Research from 2012 shows that during about half of their time online, teens are engaged with more than one activity. Given the profusion of apps since then, and the increased computing power of smartphones in the past decade, it is likely that even more online time is spent multitasking these days.[7]

All this screen time and digital communicating has been raising questions for almost twenty years now, leading researchers to study their effects on the developing adolescent brain. The findings are mixed at best and often point to concerning consequences.

One area of study, familiar from earlier chapters, is attachment. In part 1 we learned that early attachment is a strong predictor of healthy adult relationships and that screen time in childhood can have negative effects on attachment. So it is reasonable to wonder whether the same outcomes are visible in teens. A 2010 New Zealand study of over 4,000 interviews with fifteen-year-olds, some conducted in 1987 and the rest in 2004, sheds some light on the subject. Researchers measured effects on attachment by looking at correlations between adolescent screen time and the quality of parent and peer relationships. The researchers were able to learn about media-consumption effects generally and, by comparing the teenagers interviewed in 1987 with those interviewed in 2004, newer digital media–consumption effects in particular.[8]

Higher levels of media consumption were associated with lower attachment between parents and their adolescent children in both samples. In the 2004 sample, spending large amounts of time on computers and video-gaming was associated with outcomes akin to those of heavy television viewing. More digital media time also correlated with lower attachment to parents and peers. By contrast, in both the 1987 and 2004 samples, time spent reading books and doing homework was associated with the opposite outcome—improved attachment to parents.

This study was suggestive of media time's impact on relationships, but again, correlation is not causation. In addition, the interviews preceded the penetration of smartphones in the lives of teenagers. That said, more recent research continues to show apparent associations between digital technology use and issues with mental health and general well-being.

New studies are also helping to refine our understanding of the complexities of digital-media behavior. Compulsive adolescent texting and school performance is a case in point. A 2017 study defines compulsive texting as frequent texting that could not be controlled and that interfered with other tasks, such as chores and homework. Adolescents who engaged in compulsive texting felt high levels of anxiety when they were prohibited from engaging in the behavior, and compulsive texting was strongly associated with reduced academic performance. On the other hand, those who could moderate their texting behavior were able to maintain a solid level of academic performance. This suggests that some teens can in fact navigate the digital age ably. The trouble is that often teens are not especially good at self-control, thanks to their underdeveloped prefrontal cortices. And technology companies know just how to exploit impulsive and insecure young people.[9]

FOMO and the Social Media Seesaw

Jane is a fifteen-year-old girl struggling with pimples and dreaming of prom dates. And, like many fifteen-year-olds, she uses social media to communicate with friends throughout the day and often late into the night. Lately, she has been noticing that other girls from school are posting pictures of themselves on beaches in bikinis during spring break. Anxiety creeps in as she views the photos. Why wasn't she invited? And why doesn't she look like that in a swimsuit? As she stares into the mirror, her smartphone by her side, she feels she doesn't measure up. Perhaps if she just stopped eating for a month and started working out every day, she, too, could look like those girls. After a few days of starving herself, she realizes she can't go without food for very long, and she really doesn't like to exercise. She returns to her usual habits until the siren songs of social media call her back and leave her worrying, again, that she should be somewhere else or someone else.

Social comparison, cliques, and emotional swings have long been part of the adolescent experience. Body-image anxieties can be all consuming. But before the age of ubiquitous digital media and communication, it was harder to know when you were missing out or falling short of some perceived standard. The sheer quantity of information is greater now that filtered and curated images of teenage existence are posted all the time—comparisons are hard to avoid. Access to peer group behaviors and feedback is as instantaneous as it is overwhelming and constant, giving rise to new anxieties in the digital age, such as fear of missing out, or FOMO.

FOMO is a fresh term for an old phenomenon, but today's FOMO is very different from that of days gone by. In the social media era, there are few secrets left. Social media feeds are brimming with photos of parties we were not invited to and vacations we haven't taken, intensifying the anxieties associated with social exclusion. Even more insidiously, these communications are accompanied by online social metrics that did not exist in the past. Post a photo, video, or comment on social media, and the response comes with a score: how many likes and views were generated. Every one of those viewers knows Jane wasn't at the beach. And the slew of likes reinforces the desirability of those impossibly thin bikini bodies on the screen. The feedback is immediate, judgmental, and emotionally fraught. Never mind that the images probably are not faithful reflections of real life; photos on social media are often carefully curated and may be altered using software filters that blur the lines between reality and fantasy.

For Jane and millions of others, things get complicated quickly. Again, there is nothing new about social comparisons, which emerge almost inevitably as adolescents separate from parents, spend more time with peers, and explore risk-taking behaviors. But so much has changed. Victor Strasburger, a pediatrician and adolescent medicine expert, puts it this way: "Kids used to compare themselves with their peer group at school or they'd watch TV and look at Fonzie on 'Happy Days' and say, 'Hey, I'm not as cool as Henry Winkler.'" The available set of comparisons was relatively small. Now teens log on to social media and find instantaneous comparisons with limitless peer groups and socially savvy celebrities all the time.[10]

What is the impact of ubiquitous, instant digital social comparisons on the teenage brain? Lauren Sherman, a user-experience researcher previously associated with UCLA's Children's Digital Media Center, recruited teen participants who had submitted photos of various subjects as preparation for a brain-imaging study related to social media. Sherman showed the teens 140 images on a computer screen, including 40 of their own images, while their brains were scanned. There were three types of photos included in the study: the images the participants submitted; "neutral" photos, such as images of food; and "risky" photos, including images of teen alcohol use, cigarette smoking, and adolescents in provocative clothing. Each photo was displayed with a number next to it; participants were told that the number represented the number of likes the image generated among peers online. In fact, the number of likes was manipulated by the researchers so that half of the participants saw a large number with any given photo and the other half saw the same photo with the number of likes significantly reduced. The goal was to see the impact of perceived popularity on the teen brain, independent of the type of photo viewed. Participants were also asked to click on the images they especially liked.[11]

When teens saw a large number of likes on their own photos or photos from supposed peers in the made-up social network, the same reward centers active in the brain when we eat chocolate, drink alcohol, take drugs, or have sex were activated. The teens' behaviors were also highly influenced by the number of likes displayed with the photos. They were more likely to click on images, regardless of the image type, if the number of likes was high. Interestingly, when the teens looked at the photos of risky behaviors, they showed less activation in areas associated with cognitive control—including the prefrontal cortex—than they did when they were showed the neutral photos.

The conclusion? A single number next to a photo can affect the vulnerable teen brain. In addition, viewing images of risky behaviors with a large number of likes appears to simultaneously increase activity in areas of the brain associated with reward and decrease activity in areas associated with impulse control—precisely the pattern of brain activity that the teen brain is well suited to produce. When asked about the findings, Sherman suggested that online likes represent a new kind of peer pressure. "In the past, teens made their own judgments about how everyone around them was responding. When it comes to likes, there is no ambiguity."[12]

Other studies agree that teens are feeling the consequences of growing up in the digital age. A 2016 survey showed that 50 percent of teenagers feel "addicted" to their devices and 24 percent report needing near-constant connection with their phones. In a 2018 survey, more than 80 percent of 13–17 year-olds reported that social media allowed them to feel more connected with friends, but 45 percent also reported feeling overwhelmed by drama on social media, and more than 40 percent felt pressure to construct only positive images of themselves online.[13]

Other research suggests that the types of interactions teens have online matters. For example, one study of high schoolers showed that going online to communicate and strengthen bonds with "high-quality" friends has a positive role in adolescents' sense of identity. However, teens also face pressure to post photos and comments that do not reflect their core beliefs and to join "low-quality" friends in cyberbullying. The study found that these behaviors had a negative impact on teens' sense of self. A follow-up analysis of US teens in 2018 found that social media was as likely to promote feelings of closeness and affirmation as feelings of disconnection, distress, and envy, resulting in what the authors described as an emotional seesaw.[14]

An emotional seesaw indeed. The developmental gap between weak impulse control and fully engaged emotion and reward centers leaves teens particularly susceptible to outside influences, social comparisons, and risk-taking behaviors amplified by the ubiquitous use of mobile media, communications, and information technology. Teens need special help to navigate the swings of social media to avoid the worst effects while their brain centers and neural networks are still developing. In time and with the proper guidance, the prefrontal cortex and its connections to other brain regions will mature, enabling the wide range of executive functions that adults need to survive and thrive. But until brain maturity comes, tweens and teens are at particular risk in the digital age.

Even More Media Multitasking

It is not hard to imagine a fifteen-year-old in her room doing homework. She lies on her bed with a schoolbook open for studying, earbuds in and smartphone at hand. She is on social media channels with her friends, no doubt, and nothing stops her checking and posting to three or four different conversations between attempts to focus on her studies. Her parents look in on occasion and convince themselves she is studying because she looks so engaged. She is the very image of the teen multichannel multitasker, who simultaneously surfs the web, communicates online, games with ease, and does her homework.

Media multitasking has become such a common habit that few of us, at any age, realize how often we engage in it. Yet it is worthy of attention. As we have seen, media multitasking has considerable impact on cognition and academic performance among young children. The same is true of teens. Graduate students who use or have access to Facebook and text messaging while studying or during class are at higher risk of getting lower grades.[15]

There is strong evidence that the number of teens in this at-risk group is rising. Some of that evidence comes to us from Don Roberts, a communications scholar who in the 1980s carried out important studies of media depictions of violence and their impact on kids. Roberts isn't opposed to young people consuming media, though. He suggests that adolescents, curious but self-conscious, can benefit from health information and sexual advice accessed privately online. He also thinks that media has made kids today more tolerant of cultural difference and has helped socially awkward and isolated children make meaningful social connections. But he is also concerned about the potential risks of rising rates of media multitasking.[16]

Roberts coauthored two large studies on the topic, involving more than 2,000 students ages 8–18. The first, published in 2005, showed clearly that the phenomenon of youth media multitasking was emerging. Roberts and his colleagues found that children were engaged with media for 6.5 hours per day but in that time were exposed to 8.5 hours of media. In other words, more than a quarter of the time, kids were using more than one media device simultaneously. A third of the adolescents in the study reported using multiple devices "most of the time." Importantly, those who consumed more media generally were also more likely to be media multitaskers.[17]

The second study, five years later, found that seventh-to-twelfth graders had increased their media exposure significantly, in part by nearly doubling the rate of media multitasking. By 2010, tweens and teens were packing nearly eleven hours of media into eight hours a day. The earlier report had concluded, "It seems clear that media multitasking is a growing and potentially important phenomenon." Those were prescient words, as Roberts's follow-up research found. Recent findings emphasize the point. A 2017 study showed that most teens reported media use while doing homework: 50 percent indicated using social media, 51 percent watching television, 60 percent text messaging, and 76 percent listening to music.[18]

Decades of research on various types of multitasking (sometimes referred to as dual-task processing) have demonstrated that people are less effective when performing more than one task at a time. In fact, multitasking studies consistently show reduced productivity due to decreased speed and accuracy of information processing.[19] Our brains have limits, and there is no reserve we can turn to when that limit is reached. What happens when you try to do more without adding capacity to meet that demand? We work harder to achieve less.

The Brain on Media Multitasking

There are two competing theories about the brain mechanisms underlying our poor performance with multitasking in general and teenagers' very poor performance in particular. Both theories involve the prefrontal cortex, which should raise alarms for anyone who cares for or teaches teens. As discussed, the teenage prefrontal cortex is at a developmental disadvantage, lagging behind the emotion and reward centers of the brain. This lag is a factor in teens' strong tendency toward media multitasking and the distractions it creates. Media multitasking is especially detrimental to teens because it further stresses an overmatched prefrontal cortex at precisely the wrong developmental moment.

Both theories as to why we consistently fail at multitasking have to do with information management, a key aspect of executive function. The prefrontal cortex helps us sort through and direct attention to diverse incoming streams of information by recruiting different resources from across the brain as needed—that's why connectivity is as critical to cognition as regional specialization.

This is precision work: the prefrontal cortex calls on the appropriate brain resources at the appropriate moment, in order to handle tasks where milliseconds matter. An underdeveloped prefrontal cortex with immature connectivity among networks sometimes struggles to filter out distractions and direct attention even when monotasking, and the teen brain is even less effective when attempting to manage the conflicting streams of information that result from multitasking.

To understand the two competing theories of how the brain consistently fails to manage attempting more than one task at a time, we return to the metaphor of a conductor orchestrating our brain's symphony of neural activity. According to one of the theories, the prefrontal cortex attempts to recruit additional areas of the brain to handle the increased workload needed to do multiple tasks simultaneously. Imagine the conductor trying to organize an orchestra to play two competing scores of music simultaneously. If both scores call for the string section, the first theory suggests the conductor will substitute another group of instruments, such as the otherwise-idle wind section to handle the part. The wind section is not well suited to perform string parts, so the music loses some of its quality. Perhaps the harmony is off because the winds don't have the same pitch range as the strings, or the timing is ragged. Still, the winds can play most of the notes, and some of the musicians can transpose the score on the fly. A lesser conductor might have called on the percussionists to step up and try. In either case, we muddle through both tasks, but the music suffers.

The second theory of how multitasking taxes the brain suggests that we cannot and do not recruit supplemental brain areas. The prefrontal cortex simply tries to increase the demands on the same regions responsible for carrying out the tasks at hand. This creates a bottleneck, reducing speed and accuracy as task demands exceed the specific brain areas' capacity to meet them. In this scenario, the conductor is asking the strings to play two scores simultaneously. They make an awkward attempt, as the conductor struggles to gesture competing signals. The music does not sound as good, but not because the wrong musicians are being asked to play the parts. Rather, the strings do their best to handle both parts, but with diminishing returns. The maestro losses his mojo. All the parts sound worse than when played one after the next.

Mona Moisala, a Finnish psychologist with a penchant for working long hours, set out to determine which of the two theories of multitasking is correct. Moisala specializes in adolescent development and technology's effects on the brain. She is passionate about helping the world understand brain function as it relates to media. She and her colleagues sought to test both theories by using brain imaging. Their study, published in 2016, firmly suggests that the bottleneck theory is right.

The study is a two-part affair, the first focusing on adults and the second including adolescents and young adults. In the first part, Moisala's team recruited adults to read or listen to congruent sentences ("the pizza was too hot to eat") or incongruent sentences ("the pizza was too hot to sing") while lying in a brain scanner. The participants' task was to correctly identify the incongruent sentences. Participants were presented with three versions of the task, each with varying difficulty. The simple task presented sentences, one at a time, sometimes spoken and sometimes written. The harder task also presented one sentence at a time but alongside a distracting stimulus, such as background music. The hardest task presented congruent and incongruent sentences simultaneously in written and spoken form. The study design allowed Moisala and her colleagues to test both written (visual) tasks and spoken (auditory) tasks, which demand increased activity in different parts of the brain.

Consistent with prior research, the results demonstrate that information-processing accuracy—in this case, accuracy in correctly identifying the incongruent sentences—drops as distraction levels and task demands increase. The study confirmed, once again, that the more distracted we are, the lower our productivity. But Moisala's team was able to generate new insights, too, thanks to brain imaging, which clearly showed that divided attention tasks did not result in increased activity in areas of the brain outside those areas expected to be active. That is, participants did not recruit other areas of the brain to compensate for the increase in task demands. Instead, the brain scans showed increased activity in the areas relevant to the task—the visual cortex for written sentences and auditory cortex for spoken sentences. And the scans showed increased activity in areas of the brain vital to the executive function that orchestrates brain activity: the medial and lateral prefrontal cortex, which were also active when the tasks were performed separately and without distraction.

It is important to note that these brain areas showed more activity as the task got harder and accuracy decreased. This suggests that a type of interference or bottleneck in executive function occurs as the task demands increase and performance drops. It is as if the prefrontal cortex strains to meet the increased demands of working on two tasks at once but reaches a capacity limit beyond which it just cannot get the job done as effectively as it would a single task, free of distraction. The conductor is vigorously working to get the strings to play two parts simultaneously!

This first part of the research focused on language processing under conditions of distraction and multitasking. But what about media multitasking? Some have argued that growing up in a world with media multitasking should make so-called digital natives better at it. Given the ability of the brain to adapt and rewire itself in response to many environmental circumstances, one might reasonably speculate that years of multitasking could result in a generation of supertasking teens. In the second part of the research, Moisala and her colleagues decided to find out.

The researchers this time repeated part one of the study, but with a larger sample that included adolescents and young adults who reported their levels of daily media multitasking. Like other studies, this one divided the participants into heavy and light media multitaskers. The results showed that a higher level of media multitasking was associated with worse performance on the distractor task and higher activity in the prefrontal cortex: heavy media multitaskers worked harder to achieve a worse result. During the divided-attention task, the hardest task, there was no difference in performance across groups—regardless of levels of media multitasking, performance dropped. Thus, there was no obvious advantage to having spent countless adolescent hours engaged in media multitasking. The authors concluded that heavy multitaskers were more easily distracted and were not more likely to perform better on multimedia tasks. The results confirm that media multitasking not only is associated with performance costs but also that there is no benefit to it.[20]

Young people especially may do a lot of media multitasking, but they aren't gaining any useful skills from it, even as they suffer in other ways. But don't tell them that. Studies consistently show that young adults have an inflated sense of their productivity while multitasking. Young adults who engage in frequent multitasking—media or otherwise—perceive themselves to be highly effective at it. After all, they are getting more done. Right?[21]

Not really. When we multitask, we don't get more done. We just expend more effort and strain areas of our brain. The feeling of effort, in part, explains why we all multitask. We see from neuroimaging that the prefrontal cortex—which, in the teenage brain, struggles to keep up under the best conditions—works harder during multitasking, as the flow of information creates a processing bottleneck. Multitasking can also result in more effort expended across the brain, as visual tasks impose demands on the visual system and verbal tasks engage the auditory cortex. When you read while listening to music, your brain is more active than when you perform just one of these tasks. Increased brain activity occurs below our conscious awareness, of course. But in this case subjective experience dovetails with what is going on in the brain: it feels like we are working harder, because we are.

This notion that we are working harder and therefore getting more done is a trick our brains play on ourselves. On a nonconscious level, two different forces may be collaborating as part of this cognitive hoax. First, the increased brain activity relative to the task demands sends a signal to the rest of the brain that says, "Hey, I am working hard at multitasking." This signal triggers a memory association that quite reasonably says, "When I work harder, I typically get better results." Conclusion? "I must be getting better results!" Second, by activating more areas of the cortex and more functional networks, multitasking engages a broad set of neural networks, and we experience that subjectively. Our attention may be scattered while multitasking, but at least we are not bored. This is the insidious nature of multitasking, and especially media multitasking, which tends to engage both the auditory and visual cortices. We think we are being more productive because we feel more active, which reinforces multitasking behavior. But in this case, our feelings are feeding us a damaging illusion with future consequences for our mental health.

Seismic Mental Health Consequences

In 2017 Jean Twenge, a professor of psychology at San Diego State University, proposed a provocative question in an article for the *Atlantic*, "Have Smartphones Destroyed a Generation?" Twenge argues that mobile media, communications, and information technology have radically altered the behaviors and mental health of teens. She has researched American cohorts going back to the 1930s and found considerable changes over time, "not just in degree but in kind." But we don't need to go back a century to learn that the latest technology has fostered dramatic shifts. That much is apparent in a single generation.

Twenge notes major divergences in Millennials and members of Gen Z in terms of worldviews, behavior, and attitudes toward technology. While Millennials grew up with the internet, and it clearly changed the way they view the world, it was not the same constant presence in their lives as it is for the Gen Z teens. The experiences of Gen Z's teens "are radically different from those of the generation that came of age just a few years before them," she writes. Twenge describes the Gen Z teens as more comfortable inside with their phones than outside the home. High school seniors in 2015 were going out of their homes less than eighth graders did in 2009. The allure of independence that motivated previous generations to get their driver's licenses and spend time anywhere other than home seems to have been replaced by the psychological drive to hang out with friends online. Young people don't need to leave home to connect socially with friends; they just need to grab their smartphones and hide in their bedrooms.

The consequences of this change in worldview and behavior are sometimes worth celebrating. Twenge points out that the latest generation of teens uses drugs less than its predecessors and has fewer unwanted pregnancies; the teen birth rate in 2016 was down a staggering 67 percent from its 1991 peak. And teens are driving less, relying more on their parents and ride-hailing services like Uber and Lyft when transportation is needed. The result has been a statistically significant increase in teen's safety on the roads.

But there are serious costs to today's teen lifestyle as well, in particular when it comes to mental health. Twenge draws a straight line between the technological environment and teen depression. "There is compelling evidence that the devices we've placed in young people's hands are having profound effects on their lives," she writes, "and making them seriously unhappy."[22]

Important evidence of technology's contributions to teen depression comes from the Monitoring the Future Survey, a large multiyear study funded by the National Institutes of Health. Every year since 1975, the survey-takers have asked US high school students a range of questions about their health and habits. The survey has expanded over time to include new questions about the latest technology-related behaviors, but it has consistently asked teens about their happiness and other behaviors, such as how they spend their leisure time. So the survey provides a trove of comparative information that helps us tease out how life has changed for teens over the past few decades.

Recent findings show that despite being home more than their predecessors, today's teens spend less time with their families. Time spent with friends face-to-face is down, too, while more time is spent connecting with peers using smartphones and social media. The results confirm years of data showing that teenagers who spend an above-average amount of time on screen-based activities are more likely to be unhappy. The opposite is also true: those who spend more of their time on nonscreen activities, such as reading books, are more likely to report that they are happy. According to Twenge, "If you were going to give advice for a happy adolescence based on this survey, it would be straight-forward: put down the phone, turn off the laptop, and do something—anything—that does not involve a screen."[23]

We should be confident in the robustness of these findings, given their consistency over time. We should also be confident because similar results have been found in other countries with high levels of social media and smartphone penetration, which clarifies that we are not seeing the results of some other factor particular to US society. For instance, a 2017 analysis of more than 100,000 fifteen-year-old students in the United Kingdom found that high levels of screen use correlated with reduced mental well-being according to validated metrics. This study adds valuable nuance to the screen-time debate, as researchers also found that low-to-moderate levels of screen time can have a positive relationship with mental well-being. This suggests that the relationship between media and well-being is complex, as is the case with other behaviors such as alcohol consumption, that are safe and even beneficial in moderation but also subject to abuse and resulting negative consequences.[24]

These large-scale surveys are helpful in establishing a sense of the youth gestalt—whether teens feel good about their lives and whether happy teens tend to spend more or less time staring at digital screens. But the surveys don't speak to outcomes like depression and other kinds of clinically relevant mental illness. We do know that rates of teen mental illness are rising. A 2016 evaluation of data from more than 170,000 teens between 2005 and 2014 showed a nearly 30 percent increase in incidence of major depressive episodes. The increase was highest among girls, who are known to be at greater risk for depression during adolescence. The US Department of Health and Human Services reported that, in 2015, nearly one in five adolescent girls experienced at least one episode of major depression. And a 2018 World Health Organization analysis of 14,000 teens showed that more than one-third of first-year university students across eight industrialized countries reported symptoms consistent with a diagnosable mental health disorder. Tragically, a 2019 study showed that the rate of attempted suicide by overdose among US 13–18 year-olds had doubled in the previous decade.[25]

Meta-analyses suggest that screen time has something to do with this increasing incidence of depression and other forms of mental illness. And intervention studies back up this finding, helping to establish a causal link between media exposure and reduced well-being. One such study, carried out by researchers at the University of Pennsylvania, assessed whether reduced social media use affected mood. The researchers recruited 143 undergraduates and assessed their baseline mood and well-being. The students were then randomly assigned to one of two groups. The first group of participants were asked to limit their use of social media, including Facebook, Snapchat, and Instagram, to ten minutes per day throughout the three-week period of the study. The second group were asked to maintain their usual pattern of social media use. To eliminate self-reporting biases and ensure accurate assessments, the researchers used technology that allowed them to monitor the participant's use of social media apps.[26]

The study found that the group who limited their social media use had significantly lower rates of depressed mood and less reported loneliness compared to the control group that continued their typical social media use. The improvements in well-being were especially pronounced in participants who were more depressed at the start of the study. Aware of the irony of the finding that less social media use makes people feel less lonely, one of the authors speculated that "when you're not busy getting sucked into clickbait social media, you're actually spending more time on things that are more likely to make you feel better about your life."[27]

There are many risks associated with increased depression, including decreased productivity and social isolation. The most consequential is suicide. One risk factor for suicide is self-injurious behaviors, which include superficial cutting and burning of the skin, which may not lead to death but is correlated with increased risk of lethal suicide attempts in the future. While more data are needed to see how much these suicide-related behaviors are increasing, one report from California's largest school district found more than 5,000 incidents of self-injurious behaviors in 2015, compared with only 255 such incidents just a few years earlier, a nearly twentyfold increase.[28]

Mental health professionals are also raising concerns that it is often harder to diagnose depression in teens, who sometimes manifest symptoms that are less common and subtler than do adults, such as increased sleep and irritability. As a result, we may be underestimating the rate of mental health problems alongside the true impact of changing times. Unfortunately, while the incidence of depression and other mental health issues increases during adolescence, there is no commensurate increase in the availability of clinical care and treatment. Indeed, most treatments continue to be recommended only for adults, creating frustration for parents and adolescents in need.

Screen time itself is probably not all that is leading to problems of teen lone-liness and depression. What matters too is the content teens are absorbing on the screen. Writing in the *New York Times*, columnist Frank Bruni notes how society encourages "today's exhausted superkids" to strive too hard. Before the age of smartphones and social media, many parents pushed their kids to get into the best colleges by overextending themselves with extracurricular ac-tivities and slews of Advanced Placement classes. Now that pressure is being augmented by peers announcing their successes and curating their seemingly perfect lives on social media, driving an already hypercompetitive culture to new heights.[29]

The result is that teenagers increasingly experience what used to be largely an adult phenomenon: chronic stress. In addition to struggling with depression, today's overscheduled, overstimulated, and sleep-deprived teens have more anx-iety than ever. Between 2007 and 2012, the rate at which anxiety disorders were diagnosed among kids ages 6–17 increased by 20 percent, data from the Na-tional Survey of Children's Health show. Referrals for mental health disorders are increasingly common as teens enter college, according to a 2015 *Chronicle of Higher Education* report. The report shows a nearly 50 percent increase in requests for mental health counseling for undergraduates during a similar pe-riod to the one covered by the National Survey. In part the higher rate of refer-rals may be a product of reduced stigma surrounding requests for assistance, as well as increased rates of treatment during childhood, which means that more and more students are entering college with a mental health diagnosis. But those diagnoses are happening for a reason, and there is little doubt that the late teens are a difficult time in terms of managing emotions.[30]

This is not to suggest that social media–driven competition, reduced face-to-face interactions, the body image pressure of Instagram, and the destructive allure of media multitasking are entirely responsible for the increasing mental health struggles of tweens, teens, and young adults. Certainly pressure-cooker academics, the stresses of social and economic disadvantage, and an as-sortment of global crises are in part to blame as well. But given what we know about brain development in adolescence, there is good reason to worry that the overuse of smartphones and other media technologies is a significant source of the widely documented teen mental health problem.

Social media is perfectly tailored to exploit people when they are most vul-nerable: when the teen emotion and reward centers have matured ahead of the regulatory mechanisms managed by the prefrontal cortex. And it is especially easy to fall into a routine of media multitasking as adolescents, when we are most prone to social comparisons and poor impulse control. As we will see with adults, too, the connectivity between the prefrontal cortex and the subcortical areas focused on reward and emotion are intimately involved in mental illness, and we have an increasingly clear sense of the mechanisms whereby excessive exposure to mobile media, communications, and information technology can cause harm during key developmental years. Meanwhile, the case for a causal connection between social media and teen depression and anxiety is considerably strengthened by both intervention studies and surveys, including surveys conducted by social media powerhouses like Facebook. Causation is always hard to prove, but the preponderance of the evidence weighs heavily.

IN BRIEF: LOST SLEEP

One of the most important consequences of increased technology use in adolescence is sleep deprivation among a population that needs more slumber than it gets. A review of surveys covering more than 200,000 US eighth, tenth, and twelfth graders from 1991 to 2012 found a concerning increase in the number of adolescents reporting inadequate sleep. By 2012, the study found that around 55 percent of adolescents were reporting getting less than seven hours of sleep each night, while reports of adequate sleep were decreasing across all adolescent age groups. The problem of sleep deprivation among adolescents appears to be global, with findings from Germany, India, and parts of Taiwan suggesting that, on average, high school–age students in those places were getting less than eight hours of sleep. South Korean adolescents appear to be especially sleep deprived, getting on average less than five hours of sleep per night, according to a 2014 study.[31]

One reason adolescents are getting less sleep is that they are spending nighttime hours with media technology. Surveys of US 12–18 year-olds find that over 80 percent of kids engage in electronic activities after 9:00 p.m., often a combination of watching TV; using computers, smartphones, and tablets; and playing console video games. The use of television and computers before bedtime has been consistently associated with falling asleep later, difficulty falling asleep, shorter total sleep time, later wake-up times on weekends, and more daytime sleepiness. One review showed that 90 percent of relevant studies published between 1999 and 2014 found a significant relationship between screen time and sleep outcomes, with more screen time associated with delayed sleep and lower quantity of sleep.[32]

It is perfectly normal for adolescents to go to sleep later than their younger siblings. The teenage years witness a developmental shift toward later onset of sleep. Scientists describe this as a reduction in sleep pressure: our changing biology makes it harder to naturally fall asleep as we mature. But screens make falling asleep harder still. Digital devices emit predominately short-wavelength blue light that signals daytime to the brain, further reducing sleep pressure. Media consumption, whether watching television, texting on the phone, gaming, or using social media, is also physiologically arousing, further stimulating users precisely at the wrong time—right before they go to sleep.[33]

The result for teenagers is a vicious cycle, documented by research. Teens stay up later and later as sleep pressure decreases and media technology use increases. Less sleep at night leads to more difficulty waking up in the morning. This leads to the use of caffeinated energy drinks and prescription stimulants to stay awake during the day, which leads to even less sleep pressure the following night. Adolescents then try to catch up on the weekends by sleeping in, further disrupting their sleep cycles and reducing production of the natural sleep hormone melatonin.[34]

Stimulant use and sleep disruption compounded by modern media use late at night not only decrease the quantity of sleep but also disrupt sleep quality by reducing the amount of time in slow-wave or rapid-eye-movement (REM) sleep, which is critical for learning and memory formation. Studies suggest that higher-level executive functions, managed by the prefrontal cortex that is developing in critical ways during adolescence, are particularly vulnerable to sleep disruption. This sets up the adolescent for future problems with learning and regulating their emotions and natural biorhythms.[35]

Adequate sleep is vitally important for bodily functions and brain health. The Sleep Foundation recommends eight to ten hours of sleep for adolescents. Reduced sleep in young people has been linked to increased incidence of automobile accidents and substance abuse, lower school performance, obesity, and increased mental health issues. The stakes are high—a 2011 study shows that getting less than eight hours of sleep at night is associated with an almost threefold increased risk of suicide attempts in the teenage years.[36]

Researchers have also found that the earlier parents set bedtime, the lower the probability of depression and suicidal ideation, suggesting that parental intervention during adolescence can lower the risk of mental health problems while simultaneously supporting future success and healthy lifetime sleep habits. Parents and pediatricians must take care and watch not only stimulant intake—teen drug use has long been a cause of concern—but also late-night media technology use. The AAP recommends against any screen time in the hour prior to sleep. Following that recommendation is tough in an era when adolescents have screens in their rooms, often several of them. But there may be no intervention more important these days.[37]

IN BRIEF: YOUR BRAIN ON PORN

In recent years, young men have been showing up at clinics with what used to be an unusual problem for their age group: difficulty having sex. Specifically, these men complain that they are unable to consistently acquire or sustain an erection of sufficient rigidity and duration for use in sexual intercourse. The unusual thing is their age. Men in their thirties and forties have been showing up with erectile dysfunction (ED), a condition more typically associated with men over age fifty. They had something else in common: their problems started after prolonged use of online video pornography. Doctors told the afflicted to stop accessing pornographic videos online, and most of those who complied recovered normal use of their sex organs within a month or two. This was termed a sexual "reboot."

Over time, the problem has spread to even younger men. Males in their twenties and late-adolescent boys who were compulsively using video pornography online have begun to appear in doctor's offices complaining of ED. Indeed, these younger men were even worse off than the thirty- and forty-somethings who preceded them. It took them more time to recover sexual function—on average twice as much time. The issue has grown significantly, with estimated rates of ED in men under 40 rising from less than 2 percent in 2002 to between 14 and 28 percent by 2011.[38]

Why would young men and boys, with higher levels of testosterone, healthier blood vessels, and typically a shorter history of online pornography exposure, take longer to recover? Consider the generational shifts. Older men who came of age prior to the widespread availability of high-speed internet access grew up with static pornographic images in hard-to-access print magazines. They were forced to use their imaginations to fantasize about sex. When they experimented with real-life sex, their experience became richer and evolved through intimate face-to-face interactions, physical touch, and the cocreation of the joys of sex with partners.

In contrast, young males accessing pornography over a high-speed internet connection are often exposed to screen-based, scripted, professionally produced, hard-core videos of sexual fantasies that go beyond the realities of what is considered normal sex. All of that is readily available and experienced alone. When these young men then encounter a sex partner in real life, their expectations and fantasies are unfulfilled, resulting in sexual dysfunction. This is now called pornography-induced erectile dysfunction.

This is another consequence of the uniquely malleable, rapidly developing, and highly adaptable but imbalanced adolescent brain. Online exposure to pornographic video leverages the same neurobiological vulnerability in adolescents as exposure to likes on social media. The exposure not only triggers the reward centers but also turns off the prefrontal cortex.[39]

How does this work? The answer lies in the science of addiction. Like addictive substances, online pornography becomes less fulfilling the more we are exposed to it. The brain-based reward system builds a form of tolerance, so that more stimulation is needed in order to achieve the same effect that used to be more easily obtained. In the absence of stimulation, cravings appear. Tolerance and cravings prime the motivation centers and drive us to seek out more online experiences, again and again, further altering behaviors. The preoccupation and anticipation associated with the reward is mediated by dopamine neurons in the reward centers; dopamine drives the addiction and overrides the ability of the prefrontal cortex to put on the brakes—a situation that the adolescent brain is particularly vulnerable to experiencing. Brain-imaging studies of men with internet pornography–induced ED clearly show brain and behavior changes consistent with other addictions.[40]

Teens, like most of us I expect, don't think of ED as a problem afflicting their age group. But in this respect, they are behind the times. Parents, caregivers, and physicians should talk to children about the risks of addiction to online video pornography. For some families, perhaps most, this is a hard subject to talk about, but it belongs in any family media-use plan. Boys in particular are at risk for ED at a young age, and for reduced sexual pleasure and dysfunction in later years.

6

ADULTS, THERE WILL BE CONSEQUENCES

We discussed the importance of nurturing the growth of a healthy prefrontal cortex, the conductor in our own personal brain symphony, as we mature from infancy through adolescence. But what about adults? What is the impact of evolving mobile media, communications, and information technology consumption on adults and their more mature prefrontal cortices?

Like their kids, adults have been devoting more and more time to media. Ratings data from Nielsen show that, in 1975, Americans age eighteen and older spent, on average, sixteen hours per week consuming media. By 2002 that number had jumped to forty-five hours per week, and as of 2018, close to eighty. That means US adults are spending nearly half of every day using media.[1]

This shift is a result of the same powerful forces that have dramatically altered the habits of young children and adolescents. One is the convenience and ubiquity of smartphones and other mobile media devices. This enables what some industry experts call "found time for media." We can access, and even create, media content while standing in line at the store, walking down the street, on break at work, or while sitting in a car, bus, train, or airplane.

The second major change is the rise of media multitasking. The only way for anyone to consume so much media daily is to engage in more than one screen or device at a time and to introduce screens into areas of life that were once free of them. Also like their kids, adults commonly watch TV while using a mobile phone or tablet. And adults are using screens to bring work into previously work-free zones, including our bedrooms, on vacations, and in automobiles. Employers routinely expect workers to respond rapidly, at all times of day, to chats, texts, and emails. We respond while socializing with friends, working out at the gym, eating dinner, and caring for our children. And, in spite of policies intended to ensure safety on the roads, we continue to text, check email, and browse the web while driving our cars. Smartphones accompany seemingly any task, day and night.

All this media exposure and multitasking is fostering increasingly distracted, divided, and depressed societies the world over. It's distracting us from important activities and threatening relationships—even putting fully developed brains at risk. Any decline in executive function reduces our ability to pause and think before we act, encouraging the formation of unhealthy habits that in turn contribute to mental and physical health problems. Over time, even adults are vulnerable to the insidious effects of media and communications overload.

The Age of the Superstimulus

You're sitting at work, double-checking a fact for a report that's due in a few hours. This is simple research; it should take less than a minute or two. No sooner have you begun searching, though, before a tantalizing image appears at the edge of the screen, accompanied by a jaw-dropping headline. You take the bait, and the next thing you realize, ten minutes have gone by. You have forgotten the original task, but you now know five things dogs and cats can see that humans can't. How did you fall into this trap?

The answer is superstimuli. Superstimuli are neither creations of the digital age nor essential to it; they predate mobile devices, and they aren't necessary to the functioning of smartphones, social media, or the internet. But content creators have discovered that these kinds of stimuli trick both young and old brains by hyperactivating our emotion and reward centers, driving behavioral repetition and habit formation. Today many internet-related activities draw on superstimuli to deliver unending rewards designed to abuse and take advantage of our evolved motivation systems.[2]

The idea of a superstimulus was theorized in the 1950s by Nikolaas Tinbergen, a zoologist and animal-behavior specialist who would go on to win a Nobel Prize for his work. Tinbergen specialized in understanding how animals respond to visual stimuli that provoke certain instinctual behaviors. What he and his colleagues discovered was that animal instincts could often be tricked with embellished, artificial versions of natural stimuli. If a stimulus is a natural object that induces a response, a superstimulus is an unnatural object with exaggerated features that induces an exaggerated response. The animal is innately motivated to have a particular response to the natural stimulus, and that response is greatly amplified in the presence of the superstimulus. The abnormal mimic motivates the animal even more powerfully than the real thing.[3]

The discovery came in part through the study of songbird parents who instinctually feed their babies when they are hungry. Tinbergen hypothesized that the feeding instinct is triggered by the visual cue of the open, red mouths of their young. After many experiments, he learned that he could trick the songbird parents into feeding fake baby birds with highly exaggerated mouths that were wider and redder than normal. The instinct is so powerful that parents will often feed the fake birds at the expense of their own chicks. Songbirds will even abandon their own eggs, which are typically pale blue or gray in color, to sit on larger and more colorful fake ones.

Superstimuli can be made to manipulate and deceive a wide range of species. For instance, male butterflies are seduced by the colors and shapes of female butterfly wings—so much so that male butterflies will try repeatedly to mate with fake butterflies with unnaturally large and vibrant wings, even in the presence of real female butterflies. And certain species of beetles can be tricked into mating with beer bottles. Male beetles can be duped by a smooth glass bottle, brown in color, with flattened beads on it. These features resemble those of female beetles, but on a much grander scale.[4]

Besides being fake and effective, these superstimuli all have one thing in common: they are made by humans. We are surrounded by a world of superstimuli, often created in service of commercial or other gain. On a routine basis, superstimuli hijack our reward and motivation centers much like alcohol and other drugs do, provoking exaggerated responses that drive unhealthy behaviors. Food manufacturers profit by creating extra sweet and salty junk foods and highly caffeinated beverages. The web is full of altered, attention-grabbing photos of cuddly animals with extra-big eyes, to say nothing of surgically enhanced and photoshopped humans. These exaggerated stimuli tap into very old and highly evolved instincts and motivations but are derived from very new human motivations and technologies. They are increasingly paired with titillating headlines and shared on social media to enormous audiences for shock value, political advantage, or to sell products. And the stimulus they produce is so great that even a normally functioning adult prefrontal cortex, let alone an overwhelmed one, cannot easily manage them. Just as Tinbergen found in animals, for many of us, these superstimuli are difficult to ignore.[5]

Are We All Just Addicted?

Lee Seung Seop was twenty-eight years old. He worked as a boiler repairman in Taegu, one of South Korea's largest cities, and he liked to play video games. He liked playing so much that he would stay up late into the night, oversleep the next day, and miss work. Eventually he lost his job. He also lost his girlfriend, another avid gamer, over arguments about how much time he spent playing online. His family and friends grew increasingly concerned about the volume of his game-playing and the toll it was taking on his life.

One day in August 2005, he went to an internet café to play *Starcraft*, an online video game. He played for more than two days straight. When he failed to return home, his mother asked friends to look for him. They found him in the café, where moments later he passed out and was rushed by ambulance to a hospital, where he soon died.

Lee was an otherwise-healthy young man, but during the nearly fifty-hour gaming binge, something took over his brain and his body's normal functions. He was described by witnesses as having eaten very little, only taking occasional breaks to use the washroom. He neither drank nor slept for the entire time. The cause of death was exhaustion and dehydration leading to cardiac arrest. According to one friend, "He was a game addict, we all knew about it. He could not stop himself."[6]

Lee's judgment, in which the prefrontal cortex is so heavily implicated, was overwhelmed by the rewards of satisfying his gaming compulsion, to the point that he ignored the basic instincts of self-preservation. That is the depth into which addiction can drag us. Lee's death prompted global interest, for its novelty and its tragedy. Today internet addiction remains tragic, but it is no longer novel. It is a serious problem faced by many, and one does not simply grow out of it. Young people and adults alike can become dangerously hooked on digital media.

Internet Addictions Defined

Internet addictions are hard to pin down. Estimates of their prevalence vary widely, and it seems clear that incidence is mediated by culture and geography, so that the rates of internet addiction will be more common in some places than others. A 2014 review of studies indicated that as little as 1 percent or as much as 18 percent of the global population had an unhealthy relationship with internet gaming, shopping, pornography, social media, or some combination of these. At the time, the problem appeared to be most severe in Asia and the Middle East. By 2021 things looked even grimmer. An analysis just of social media use across thirty-two countries put the global range higher, estimating that 5 to 25 percent of users have an unhealthy relationship with online media, depending on the definition of unhealthy use and controlling for cultural factors.[7]

One reason that it's hard to say exactly how many people are wrestling with worrisome online behaviors is that experts have struggled to come up with a common definition of internet addiction. Often researchers will look for "problematic" use—use that causes some kind of abnormal or undesirable outcome. But sometimes the defining criterion is the amount of time spent online, so that researchers may classify "excessive" use as addiction, irrespective of whether the user or others suffer in some concrete way. However, as we have seen, apparent time spent can be an unreliable metric, because usage is easily hidden. Researchers may also look for "compulsive" use and "dependency," in which case addiction is measured according to a person's inability to resist the behavior in question.

As a result, there is no clear consensus about how to screen people for internet addiction. Complicating the debate over how to characterize and diagnose the disorder are varying cultural attitudes toward technology: what looks like internet addiction in Tokyo might register differently in Tuscaloosa, and vice versa. There are also many types of online behaviors that may be problematic; some research will focus on one type and ignore others.

Even authoritative bodies like the American Psychiatric Association (APA) have struggled to provide clinical guidance concerning online addictions. The APA's *Diagnostic and Statistical Manual of Mental Disorders* (*DSM*), the fifth and most recent edition, which was published in 2013, describes internet gaming disorder (IGD) as an addiction, albeit one requiring "additional research and evaluation." And the *DSM* does not include internet addiction more generally.[8]

The *DSM* is the key reference work in American psychiatry and is assembled by a panel of experts on the various topics it covers. After considerable review of the available research, and after much debate, the APA chose not to incorporate the broader category of internet addiction, which would include a variety of behaviors, and instead focused narrowly on gaming addiction. The rationale was that IGDs explained most of the internet addictions seen in the literature published globally until 2013. The *DSM* does not deny that other online behaviors can be problematic or even addicting, but gaming disorder clearly stood out as the one with the most documented cases when the fifth edition was under development.

What moved the panel was, in part, evidence that problem gamers behave like pathological gamblers and people with drug and alcohol use disorders. IGD isn't just another word for heavy gaming; you can be passionately devoted to online gaming without being addicted. Instead, according to the *DSM*, IGD is diagnosed in heavy gamers who play compulsively and to the exclusion of other interests and needs—up to and including the need for food and water, as in Lee's extreme case. People with this condition endanger their jobs, school performance, health, and social relationships. Importantly, they often experience a form of withdrawal when they don't play or are kept from playing for more than very short periods of time.

Additional compelling evidence comes from imaging studies that show that extreme gaming does to the brains of gaming addicts what extreme drug use does to the brains of drug addicts. We know that the brains of people addicted to drugs or alcohol are both structurally and functionally different from those of people of similar age who are not addicted to drugs or alcohol. When we review the structure and function of the brains of people with severe online gaming disorders, several studies show that they resemble those of people with drug and alcohol addictions. As the APA puts it, "Gaming prompts a neurological response that influences feelings of pleasure and reward, and the result, in the extreme, is manifested as addictive behavior."[9]

Still, the *DSM* doesn't include IGD as a formal diagnosable mental illness. This is because there remains a lot of debate over the exact nature of this condition. As noted, while there have been many studies of gaming addiction, they don't always use the same research method or definition of addiction, which makes it hard to analyze data across studies. Relatedly, different studies have different criteria for assessing whether a person is addicted, so that the stringency of the criteria have a lot to do with assessments of prevalence. For instance, some studies identify as addicted only those who engage in deception to conceal play time, producing a low prevalence result. Many more people experience withdrawal when compulsive use is interrupted, but they won't be identified as addicted in a study that focuses on deceptive behavior. In addition, some have argued that the research available so far may not be applicable across cultures and geographies. A high percentage of the internet-addiction research to date has been conducted in Asia, where particular attitudes toward technology and cultural norms surrounding social stigma may play a role in driving some extreme behaviors. The compelling brain-imaging studies are also mostly from Asian countries and are yet to be replicated elsewhere, limiting their generalizability.[10]

Another interesting piece of research reverse engineers the APA guidance with counterintuitive results, suggesting that we may not yet have the right set of criteria to gauge IGD. The researchers surveyed people in the United States, Canada, the United Kingdom, and Germany and found the overall prevalence of IGD using the APA's strict criteria to be surprisingly low: around 0.5 percent among young adults and 1 percent among adults. The reason this is surprising is that we know gaming is occupying more and more of people's time and that sales of all types of video games have skyrocketed over the last decade. By contrast, there has been a clear correlation over time between cigarette sales and cancer rates. Critics of IGD's inclusion in the *DSM* have argued that if indeed video gaming is an independent cause of addiction, we should see the number of addicts rising in tandem with the popularity of gaming.[11]

Game sales and observed addiction rates may be decoupled for a number of reasons, even as increasing numbers of people really are addicted. For one thing, gaming addiction may be easier to hide than cigarette smoking, alcoholism, or shooting heroin, because so much video gaming is done at home using mobile platforms, desktop computers, and televisions that also serve other purposes. In addition, modern internet gaming diverges from other addicting activities in an important way, which affects how we evaluate who qualifies as addicted. Specifically, whereas addictions are often diagnosed in part on the basis of displacing relationships—that is, the behavior undermines social engagement, causing the addict to become isolated and lonely—online gaming can be a social experience. The effects may be subtle. Gaming addiction may cause a person to lose friends and grow distant from family, but in the course of gaming, the player interacts with people online, forging new relationships that may be genuinely satisfying. So if a survey-taker asks heavy gamers whether they have fewer friends than their peers or fewer friends than they had in the time before they played, the answer may be an honest no.

And, as ever, there are problems of causality, which are magnified in an area of research that is relatively young: there just hasn't been enough time yet to do similarly designed long-term studies of problematic gaming behavior across cultural contexts. Research suggests that computer games both attract those who are distractible and add to the distractibility of those who use them, so it can be hard to say in any individual case whether gaming has harmed a person or whether gaming is just one of many activities that might obsess a distractible person, undermining their ability to thrive.[12]

This also reminds us that brain rewiring can work both ways: some people are wired to become heavy gamers even with limited exposure, and some start out with normal functioning but fall into pathology as repeated behaviors induce new brain wiring that reinforces gaming over and over to an extreme. Things get complicated quickly, but additional research will help us better understand problematic internet gaming—what it does to people's welfare, how it affects the brain, and how it can be addressed therapeutically.

Of course, you don't have to be an avid gamer to spend a lot of time immersed in media, communications, and information technology. Other behaviors including social media use and online shopping can absorb huge quantities of time, money, energy, and emotional resources. And remember those figures the chapter began with: adults spend almost half of their lives consuming media. Are we all just addicted?

Addiction and the Prefrontal Cortex

We are not. We're not all addicted to our smartphones or our screens in general. But most of us have picked up a range of digital habits that have changed considerably our lives and our brains. Habits so powerful that, if we are not careful, could become addictions.

Habits and addictions are not the same, but they are related. Both rewire the brain in meaningful ways that neuroscientists and addiction specialists now understand relatively well. Both habits and addictions are influenced by our brains' dopamine-based reward systems. People tend to repeat those behaviors that yield positive rewards, and the more rewarding a behavior, the more repetition occurs, increasing the risk of addiction. But habits do not inevitably spiral into addictions. Keeping the distinct processes and consequences of habits and addictions in mind is key to establishing a useful digital literacy.

How exactly does addiction rewire the brain? Well, let's say you come across a positive reward stimulus for the first time. A positive reward stimulus is anything that generates positive learning (or reinforcement) and induces what's called an approach behavior (as opposed to withdrawal behaviors induced by negative rewards). A positive reward often evokes desirable emotions such as joy and pleasure, in contrast to negative emotions evoked by negative rewards.

The brain's reward centers are primitive, having arisen relatively early in our evolution. Located in the subcortex, the reward centers have been finely tuned across eons to respond to stimuli by signaling specialized neurons to release dopamine and other neurochemicals that influence our moods, feelings, motivations, and ultimately behaviors.

Dopamine is one of the brain's most important signaling neurotransmitters. When we get enough of it, we can feel pleasure; when we get too little, we can feel the opposite. Crucially, the value of the dopamine reward—whether it makes us feel good or bad, and to what extent—is a function of expectations. The brain experiences positive rewards as a "prediction error" in our favor: we were expecting a particular dopamine release but got more. A negative reward is experienced as a lower-than-expected dopamine release. The size of the gap between expectation and experience determines the strength of the dopamine signal, and other areas of the brain interpret whether that signal is positive or negative on the basis of past experience and the context in which the action takes place.[13]

The reward centers respond to all sorts of stimuli by sending dopamine signals to multiple parts of the brain, including the prefrontal cortex, where these signals provide instructions to pay attention or prepare for one or another behavior. Dopamine signals to the prefrontal cortex and certain other brain regions also play an important role in learning and memory. In this way, the neurons that carry dopamine help train us to feel positively and negatively about certain experiences. Lots of positive reward signals keep us returning to a particular experience, while negative signals leave us dreading that experience and help us to steer clear in the future. (Dopamine is also essential to multiple brain activities apart from those that cause us to feel pleasure and pain.[14])

Beyond positive and negative stimuli, addiction experts also tend to break down reward stimuli along another dimension: natural versus nonnatural. Examples of natural rewards with a positive valence include watching a beautiful sunset, planting flowers, exercising, and engaging in enjoyable social interactions such as watching a child achieve a milestone, laughing with a good friend, or having sex. Natural rewards with a negative valence arise when we experience fear and anxiety in response, for example, to a risky situation that could cause us physical harm or a social interaction that results in disappointment. Although these experiences aren't fun, they teach us to avoid scenarios that are likely to be harmful or cause pain in the future.

This brain process repeats continuously throughout our lives. Rewards train the subcortex's reward, emotion, and motivation centers to assess the value of stimuli in our environment. Signals from these subregions are then processed in the prefrontal cortex to help focus attention and enhance learning for future responses. To say that this critical circuit—from attention to emotional experience to reward to learning and behavior—has influenced our species' survival and impact on the planet would be an understatement.

Not all stimuli that the brain perceives as positive are good for us. In contrast to natural positive stimuli, nonnatural positive stimuli, like superstimuli, are often manufactured or manipulated by humans in an effort to trigger a powerful positive reward that keeps us coming back, even though aspects of the stimulus may be harmful. There are many familiar addicting substances that trigger a powerful approach reward, including caffeine, nicotine, marijuana, alcohol, cocaine, and heroin. However, currently the *DSM* recognizes only one behavior, as opposed to substance, that is classified as addicting: pathological gambling. As discussed, addictions experts are increasingly recognizing a variety of online behaviors as potentially addicting as well.

One of the major discoveries in addiction medicine over the last half century was that all of the various kinds of rewarding experiences are processed via a common brain pathway that includes the dopamine-based reward system. It does not matter whether the stimulus is natural or nonnatural; whether it is behavioral, like watching a sunset, or chemical, like drinking alcohol. Eating chocolate has been shown to trigger the reward centers in the brain in ways similar to drugs of abuse, hence there is something to the "chocoholic" moniker. And as we saw in the last chapter, a visual cue such as a number next to a photo suggesting social media likes triggers similar activity in the reward centers of the brain. That's one of the ways Facebook and other platforms feed us nonnatural rewards that influence our feelings and our behaviors.[15]

This similarity in mechanisms of action helps us understand why addictive drugs work as well as they do. In effect, they muscle out other stimuli by being extremely efficient in the reward department: drugs of abuse trigger brief but exaggerated dopamine releases in our reward system. Scientists estimate that various addicting drugs increase dopamine levels in the reward centers by as much as two to ten times more than natural rewards, depending on the drug and the natural comparison. If the dopamine signal is sufficiently strong, the resulting brain response can be a sensation of euphoria, explaining the high we get from drugs. Eating chocolate may trigger the same positive dopamine-based reward process as injecting heroin, but the prediction error is much larger in the case of heroin, so the dopamine signal is much stronger.[16]

Another recent discovery, and one that helps to differentiate addicts from those not addicted, is that the timing of dopamine release changes as we become habituated to certain behaviors. The initial interaction with a rewarding stimulus—say, a first kiss or first hit of a drug—causes a sharp, transient surge of dopamine. The amount released correlates with our expectations of the experience and, in the case of substances, dosage and potency (e.g., cocaine is stronger than nicotine, which is stronger than chocolate). These sharp, brief bursts continue while we learn from experience. They are a necessary condition for addiction to take hold, but not a sufficient one.[17]

As we become used to what were once new experiences, we learn to predict what is coming. Over time we associate the stimulus with the reward, so that dopamine is released in anticipation of the experience, sending strong messages to other key areas of the brain related to motivation and memory. Thus, the dopamine-based reward system becomes a prediction engine that guides our behavior toward positive experiences over and over again. This is how addictions develop.

The risk for developing an addiction is tied to the valence of the reward (positive or negative reinforcement), its potency (high versus low), and its availability (easy or difficult to obtain). Risk is also influenced by the frequency of exposure and by our ability to learn, recall, and understand the dangers. The potency of a reward varies depending on a number of factors unrelated to the stimulus itself, which is why some people are more easily addicted to certain experiences than others are. These factors include the genetics of the individual, stress levels before, during, and after the experience, and the social context at the time of the experience.

A healthy and resilient prefrontal cortex can moderate the potency of the reward response. The prefrontal cortex helps to manage and interpret the information implicit in a given dopamine release: while the reward center is sending dopamine signals that may encourage us to seek out the stimulus, the prefrontal cortex is shaping and contextualizing what just happened, perhaps reminding us of the downsides and helping us reflect on future consequences. Eating chocolate feels great, our reward centers say, but the prefrontal cortex cuts in with a reminder that eating too much of it isn't healthy. Again, the prefrontal cortex is the brake that offsets the gas of excessive unhealthy rewards.

With repeated exposure—and without the intervention of a healthy prefrontal cortex—the relevant areas of the brain adjust to the reward experience by down-regulating or decreasing dopamine sensitivity. Recall that positive and negative rewards aren't a result of the absolute amount of dopamine released but rather of the gap between expectations and experience. Decreasing sensitivity means raising expectations, so that more dopamine must be released in order to foster the expectations gap that led to the pleasures of earlier times. This is how we build tolerance. To obtain the reward experience we want, we must increase the intensity or frequency of the behavior that stimulates it.

The more powerful the reward and the weaker our prefrontal cortex, the faster tolerance builds. For example, the reward centers quickly develop tolerance to opioids. After just a few days or weeks of repeated use, the positive experience wears off so that users are left taking the drug just to relieve the negative experiences of being without it, including withdrawal symptoms and cravings. The desire to "feel normal" becomes a motivation for the addicted person to acquire the reward experience. This is known as dependence. Where the rewards of the stimulus are weaker, as in the case of alcohol, it can take years for dependence to take hold.

Early on the path to addiction our prefrontal cortex is fully engaged and often aware of what is going on. This gives us the illusion of control. "I got this," or "Don't worry, Doc, it's not a problem," are common refrains I have heard treating patients who are in the early and middle stages of addiction. Yet we often lose control over time. The behavioral drive to seek out the addicting substance or experience repeatedly disrupts our previous experience of natural rewards and begins to overwhelm attempts by the prefrontal cortex to stop the addicting behavior.

Eventually the addicted brain no longer seeks pleasure from weaker natural rewards—rewards that were previously motivating and continue to be healthy in the long run but are often unable to meet the new expectations set by the addicting substance or superstimulus. Instead the brain focuses on repeating the addictive behavior in order to satisfy its dependency. Less potent natural rewards, such as eating a nice meal, being with loved ones, or watching a sunset, become much less rewarding in the face of addiction because the neurochemistry of the reward centers and the prefrontal cortex have been rewired. Physical drug dependence hijacks the prefrontal cortex, reducing impulse control and the ability to remind ourselves of the consequences of our behaviors. The conductor is left to direct the orchestra with both arms tied behind her back.

What follows is altered decision-making that is biased toward the rewards of the addicting substance or behavior, more negative moods in the absence of that substance or behavior, and a reduction in pleasure from other previously rewarding experiences. This leads us to risk significant loss of friends, loved ones, and work. We ignore once-valued experiences, as the balance of power shifts away from our prefrontal cortex to the reward centers. As the addiction gets worse, we will violate social norms and personal values in search of the now-dominant reward. That might be a drug high or the high of online gaming or social media. When the reward is a drug high, the last stage can be overdose, when an addicted person's tolerance is so high that the only pleasurable experience is a dangerous one.

In Lee Seung Seop's case, it was the reward that came from playing *Starcraft* for an hour or two that could no longer satisfy his growing dependency. Scientists can and should debate the right ways to think about online addictions. But there is no question that Lee's prefrontal cortex lost the battle with his reward system, and he became addicted. His story is all too familiar to anyone who knows that struggle. Friends and loved ones realized he was playing too much. He suffered consequences, losing a partner and a job because his behaviors were out of control. We can speculate that such frustrations in his real life contributed to his sense that only in the futuristic fantasy world of *Starcraft* could he be happy. So what started as a gaming habit evolved and eventually went from bad to worse. He indulged himself in a repeated behavior that was already hurting him, until he could no longer judge accurately what his mind and body really needed, and he ultimately lost the game of life.

Habits versus Addictions

It starts innocently. We begin to bite our nails when we are nervous. We grab a doughnut or sweet snack when stressed out because our reward centers "like" the sugar rush, and the pleasure distracts us from our problems, if only for a moment. We check our smartphones and social media when we are bored, in anticipation of a quick reward—a thumbs-up, funny clip, or titillating piece of news—to relieve our doldrums. We obsessively check our phones so we never miss a work message. Then the behavior repeats. And repeats. And repeats.

These sorts of behaviors are extremely common; in particular, turning to smartphones in moments of boredom is practically universal in societies with high penetration of mobile digital technologies. At the very least, these behaviors are far more common than addiction rates predict. In other words, they are habits. Elements of these behaviors may be related to, or precursors of, addictions, but we aren't all addicted to our devices and don't have to be in order to experience the consequences of unhealthy habits.

What is the difference between a habit and an addiction? Both involve repeated behaviors. Both operate largely below our conscious awareness. Both can have significant consequences. It can be hard to draw the line behaviorally and biologically. Indeed, some neuroscientists place habits and addictions on a continuum. Others disagree, arguing that these two behavioral paradigms should be understood as distinct in the brain. What most experts do agree on is that the relationship between two brain regions in particular, the prefrontal cortex and the reward system, stand between habit and addiction. Another key player is the striatum.

The striatum, a subcortical region that sits between the prefrontal cortex and reward centers, is involved in voluntary motor control, action planning, and reward. The striatum has a remarkable ability to combine discrete behaviors into a single, seamless whole, enabling us to execute complex actions with little conscious awareness. The operation of the striatum is analogous to that of a computer program that executes many sequences of software code all at once: all those commands are operationalized with a single keystroke. In a remarkable example of cognitive economy, the striatum helps to free up our conscious awareness to focus on other things while we execute complex behaviors without even thinking about them.[18]

As an example, recall the first time you tried to drive a car. You were consciously aware of the many steps involved: place your foot on the brake, turn the key, check the engine lights, check the mirrors, and then put the car in gear. This was the prefrontal cortex talking to the striatum: you were relying on a conscious form of short-term memory to plan and execute each discrete step in the sequence. This process requires a lot of cognitive effort. But over time, and without our conscious awareness, the striatum pulls these steps together into a single action unit, so that eventually we don't have to rely on the capacities of the prefrontal cortex in order to start the car. By creating a shortcut, an automatic behavior, the striatum frees the prefrontal cortex to focus on other things.

Automatic behaviors are the essence of habits. We don't have to think about how to do them; the striatum has compiled a behavioral program that operates without conscious direction. We may not even be aware that we are taking these actions, but we do them anyway because the striatum's other function—the integration of reward signals—has reinforced the behavior, driving repetition. But because most habits are not terribly rewarding, it is relatively easy for the prefrontal cortex to keep them in check.

So, in the case of habits, our neurobiology is in a form of equilibrium. The reward centers reinforce the experience, encouraging repetition and learning. At the same time, the prefrontal cortex moderates the experience of the reward, so that there is just enough reinforcement but not too much awareness. As a result, we can override weak habits when we need to. The prefrontal cortex performs its monitoring function, the striatum performs its learning and action functions, and the reward centers perform their motivation function. Everything is in balance

However, as habits become more routine, they are subject to less monitoring from our prefrontal cortex, and the balance of power can shift to the reward centers. This is where the line between habit and addiction blurs. On the other side of that line, fuzzy though it may be, the reward centers have become the driving force. The striatum is still working in service of the reward, even as monitoring and regulation by the prefrontal cortex are diminished or absent. At this point, the reward systems are doing their job, signaling the strength of positive and negative inputs from the environment. But with the prefrontal cortex disengaged, judgment falls by the wayside. Positive rewards still feel good, and negative rewards still feel bad, but no systems are in place to parse the larger meaning of these rewards, to evaluate the potential long-term consequences, and put the brakes on. This is how habit turns to addiction in the brain. We can tell what feels good but can no longer judge accurately whether what feels good is good for us. Even in situations where we are aware that a behavior is out of control, the primacy of the reward system over the striatum makes it hard for us to stop.[19]

It is important to keep in mind that indulging a habit is not a certain route to addiction. Many habits are not sufficiently rewarding to become addicting. In addition, other factors such as genetics, mental illness, early childhood experiences, and stress play a role in our propensity to turn a habit into an addiction. And the environment is a critical factor because it influences what rewards are available to us and whether these rewards are actually good for us or instead trick our brains. Those of us lucky to have strong relationships, fulfilling work and hobbies, the independence that accompanies physical health, and the financial well-being that enables access to multiple sources of comfort, entertainment, and intellectual stimulation will find in life a panoply of natural rewards. These social goods are no guarantee against addiction, but they significantly reduce the risk.

At the same time, we all face the risk factor that is modern mobile media, communications, and information technology. We have willingly introduced these addiction vectors into our lives. Their addiction potential may be relatively low, or it may be higher than researchers realize, given the difficulties of defining and studying internet-addiction behaviors. For Lee Seung Seop, the scientific debates are immaterial. At some point along his journey with internet gaming, his prefrontal cortex lost its ability to monitor and control his actions; his striatum built master skills at gaming that could be executed with little conscious thought, further undermining his ability to intervene against self-destructive behavior; and his reward centers propelled his habit into a deadly addiction.

Context Matters

It is worth spending a little more time on the relationship between context and addiction, which speaks to important differences in addictions to drugs of abuse and to mobile technology and media. Certainly addiction to drugs of abuse is more dangerous, on the whole, than are addictions to video games, social media, and online pornography. But, considering that we have almost universally invited always-on media, communications, and information technology into our lives, our level of exposure is extremely high, increasing the likelihood that any given person will develop problematic behaviors. And we know that unhealthy habits can, over time, have serious consequences. As we will see, the dangers associated with our smartphones can actually be quite considerable, especially when we find ourselves unable to put those smartphones down in a context where use is socially unwelcome or physically unsafe.

One of the most powerful illustrations of the influence of context on addiction comes from the US war in Vietnam. In May 1971 two members of Congress visited soldiers on the battlefield and came back to Washington with the terrifying news that an estimated 15 percent of US servicemembers were addicted to heroin or opium, which were cheap, potent, and plentiful in Vietnam. Officials and the public were understandably spooked, not least because these soldiers would, eventually, be returning home. How would the country deal with all these wounded warriors? And just how bad was the problem, really?

President Nixon responded quickly by establishing a Special Action Office of Drug Abuse Prevention to study the situation. The office's administrator, addiction expert Jerome Jaffe, arranged for every returning soldier to be urine-sampled and drug-tested in Vietnam before going home. Those who tested positive were kept in Vietnam until they detoxed and only then came back to the United States. Jaffe hired a psychiatrist, Lee Robins, to lead a follow-up study to compare the detoxed returnees to a group of returnees who tested negative. Hopes for the soldiers who tested positive were not high. For one thing, it turned out that the opioid problem was even more widespread than feared: Robins found that 19 percent of a large sample of soldiers had become users. Given the powerfully addicting nature of opioids and the substantial relapse rates among heroin addicts in the United States and elsewhere, the assumption was that few of the soldiers would be clean for long. Robins followed their progress to find out.

The results were so remarkable that hardly anyone believed them: only about 5 percent of the returning heroin users relapsed within the first year. When the results were released, many experts dismissed the finding as impossible. But over time the study became widely recognized as an important contribution to our understanding of the powerful role of environment in addiction. Of the soldiers who relapsed, many had social problems prior to serving in Vietnam or had issues with other drugs while there, suggesting a compromised prefrontal cortex and reward system. But in many more cases, where these other risk factors did not exist, it turned out that once the soldiers were removed from the stress of combat and from an environment where narcotics were widely available—at the time, opium, heroin, and opioid pain medications were not easily obtainable in the United States—the reward systems in their brains received far fewer cues that drove cravings and use.[20]

Today we have a much better understanding of the variables that affect addiction and relapse rates. Hardly anyone disputes that context is a key driver of habits and addictions as well as the circumstances under which the former become the latter. Importantly, we have come to understand that as the brain experiences any reward, it learns to associate that reward with contextual cues. When these cues arise, the reward centers in addicted people's brains are triggered in anticipation of the experience. In the absence of contextual cues, little or no dopamine is released in anticipation of the reward because the brain thinks the reward is not available. As Jaffe put it to an interviewer some forty years later, "I think that most people accept that the change in the environment . . . makes it plausible that the addiction rate would be that much lower."[21]

Now consider our habits around mobile media, communications, and information technology, including media multitasking. We tend to form these habits in relatively innocuous situations, for example by watching TV and using our smartphone at the same time. Or we build the habit at work, checking our phones repeatedly while doing a separate task on a computer. In these cases, the consequences are minor: a missed moment of a TV show or lower productivity at work.

But then we get into our cars with our smartphone by our side. In this environment, the phone's unconscious triggering of our reward centers can have much more serious impact. Our reward centers still respond impulsively to our phones, averting our eyes from the road when they call to us. The prefrontal cortex may not be able to check in. Recall that multitasking, like driving while using a phone, is an automatic habit that can overwhelm the brain with obligations. The dual task of driving and using our smartphones taxes the coordinating and monitoring functions of the prefrontal cortex. If the conductor does check in, he may be too late to prevent a seriously wrong note.

The key question here is why the brain barely registers a difference between these two scenarios. Why does smartphone multitasking in the safety of our home strike the brain as similar to multitasking on the road, where such behavior is, as discussed below and in chapter 9, a recipe for disaster? The answer is that context matters. We have evolved to respond to environmental cues that trigger rewards. These cues helped us form habits that were essential for our survival, habits that enabled us to navigate the world, find food, and form social relations without needlessly burdening the prefrontal cortex. But this brain mechanism can be tricked by human-constructed stimuli.

The same network of neurons that drives our responses to natural rewards becomes hijacked as we try to navigate the road because our smartphone is its *own* context. It creates a familiar environment wherever it goes. It has its own unique and varied cues—a particular shape, lights, sounds—that send small but powerful signals to our brains; it provides moments of reward that drive our behaviors. And for most of us, the phone is there all the time, an arm's reach away. We cannot leave the context created by our smartphone the way US soldiers left Vietnam.

So is this habit or addiction? In the moment when we are driving and using our smartphones, our brain has been tricked in much the same way that a drug of abuse can trick us. The task demands of driving a car are much higher than those of watching television in our living room or working on our computers.

The sequence of behaviors needed to drive a car has become so habituated in our minds that we forget how complicated it is. Using our smartphone is also a firmly ingrained habit, so this task also seems easy. Our striatum is doing most of the work on a nonconscious level, while our prefrontal cortex is compromised by significant workloads and a heavy form of multitasking.

What do you call it when you are powerless to prevent yourself doing something rewarding, against your better judgment? What about when that behavior can have deadly consequences, as is the case with the rising number of automobile accidents attributed to smartphone use in cars? Life consequences are often seen as a key difference between habits and addictions. An inability to stop a behavior despite trying is another. Risk-taking is a third. We know that using the phone while driving is risky, and it seems clear that most of us would like to prevent this behavior in ourselves but struggle to do so, hence the popularity of technological fixes: special apps, phone settings, and in-car technology that are designed to reduce distractions and make smartphones harder to use while driving, to protect us from the powerful impulses we have developed. Some experts would argue that when we cannot control impulses that pose elevated risk of serious consequences to ourselves or others, it is time to describe that behavior in terms of addiction.[22]

Whether or not technology multitasking while driving is a sign of addiction, few would disagree that it is at least an unhealthy habit. Unhealthy habits, like all habits, arise from behaviors that change our brains—the rewiring of this book's title. The positive rewards that accompany certain behaviors train our brains to become hyperefficient in carrying out those behaviors— efficiency that is achieved by removing our conscious minds and our prefrontal cortex from the process. With sufficient rewiring, the behavior becomes automatic. This rewiring can come in response to healthy and unhealthy behaviors, either of which may present positively to our reward centers. This is the blessing and the curse of the evolutionary marvels inside our heads. Our brains "want" positive rewards and will pursue them whether or not they are good for us.

Other Cognitive Consequences

What is so insidious about smartphones and associated social media and other applications is that, like designer drugs, they are constructed with the explicit purpose of dosing the brain with positive rewards—lots and lots of them—to build habits of use, largely for the economic gain of others. It is not by accident that these technologies rewire our brains—that is what they have been designed to do. Where consequences are severe, addiction is probably the right framework for understanding the problem. But even where the effects are not as extreme, these habits, newly wired into our brains, often have concerning consequences. Let's explore a few.

The Google Effect

Modern computers and the internet were supposed to make us smarter. The idea was to augment our human mental capabilities, not supplant them to the point of diminishing returns.

From my own experience, I'm not sure I can say it worked out so well. As the digital age raced along, I began to get a sense that I was processing information differently, but not for the better. Even as I was beginning to gather sources and ideas for this book, I noticed that I was skimming short articles and online headlines more than reading them. Academic journals and printed books began to feel intolerably long; instead, a search engine could find summaries of complex topics that I could read quickly and without much effort. I did not bother to memorize key findings, because I knew I could search for them later. As time went by and I felt the pull of these new behaviors, I began to wonder whether my technology habits were making me less intelligent. Is this possible? Let's look at the evidence.

Signs of trouble predated the proliferation of the smartphone. Around the turn of the twenty-first century, as the World Wide Web was becoming widely popular, researchers were becoming increasingly concerned about the power of internet search engines to displace some of our intelligence. The worry was that push-button access to information would degrade our brains' memory systems because there would be less need to store that information ourselves. It took a little time, but in 2011 researchers published a series of studies that strongly suggested this hypothesis was correct. Access to information online was altering the way we store information in our brains, with very real cognitive consequences.[23]

One of the experiments looked at the concept of directed learning. Directed learning is the ability to store information in our memory more efficiently when we believe we will need that information in the future. For example, in order to ace a test or complete a work assignment, we will work harder to remember relevant information and will therefore more likely recall that information. For purposes of the directed-learning experiment, participants were asked to use a computer search engine to find answers to trivia questions. Half the participants were told that the computer would store the results and half were told that the results would be automatically deleted. Consistent with prior research, those who believed the computer would store the results for future access had worse recall of the information.

In other experiments, participants had a better memory for where files containing factual statements were saved on a computer than for the statements themselves. This suggests that knowing that information will be readily available in the future alters the way we encode or store memories in our brains. This is a shift toward cognitive economy; it may be less taxing to remember where to find certain information than to remember the information itself. In follow-up research, the authors found that most study participants did not feel the same anxieties I did when presented with the prospect of offloading mental functions to a computer: having online access to trivia answers made people feel smarter when compared with people who lacked that access.[24]

The researchers concluded that there are cognitive consequences to having information freely available at our fingertips. They called these consequences the Google effect. The Google effect is a kind of internet-induced amnesia: the tendency to more readily forget information that we think we will be able to find easily online. Ironically, becoming forgetful in this way—and therefore more reliant on computers rather than our own brains—makes some of us think we are getting smarter.

Overworking Our Working Memory

You are driving to meet friends at a new restaurant in an unfamiliar part of town. Your smartphone battery ran out, so you can't use the GPS to find your way. Instead, you do something you haven't done in ages: you pull over and ask a stranger for directions. Fortunately, this stranger knows just where to go. "Head straight up this road, turn left at the bike shop, then carry on for two blocks until you see a small park, and the restaurant is on the right," she says. You are anxious because you are late, and after turning right at the bike shop, you lose your way again. You forgot the rest of the directions. No worries, you can just call the restaurant—except your phone is dead.

You have just experienced a technology-enabled failure of short-term memory. Short-term memory, also known as working memory, is the ability to hold small bits of information in our brain for immediate processing. Think of working memory as the scratch pad of consciousness used for immediate needs, whereas long-term memory is the filing system where we store information for future retrieval. The Google effect is one kind of memory deficit induced by information technology, in which the information from the scratch pad never makes it to the filing system. Our hypothetical diner is suffering working memory failure due to anxiety—abetted, in this case, by dependency on technology.

The details of working memory, and working memory failure, constitute an active area of research. Much has been learned over the past few decades. For instance, we have learned that different senses, such as vision and hearing, are associated with different working memory capacity. Let's focus on the visual system. Think of visual working memory as a set of cubbies where we temporarily store the objects we're currently seeing, while our brain decides whether they are worthy of additional processing and storage in long-term memory. Most people can process three or four visual objects simultaneously in their working memory, though the number varies. The ability to process visual stimuli in working memory is tied to a variety of cognitive skills. For instance, we need to be able to filter out irrelevant objects in our visual field and focus on relevant ones in order to undertake spatial navigation. Greater visual memory has also been linked to a particular type of intelligence called fluid intelligence: the ability to solve novel problems independent of knowledge from the past.[25]

Memory experts have long understood that the brain has limitations and makes trade-offs between information processing and information storage, so that when we are working harder to parse multiple inputs, we have less capacity to store new information in working memory. (This is one of the ways in which the brain is not like the computers we build: in the brain, memory capacity is not entirely distinct from processing power, whereas in a digital computer, computation is segregated from data storage.) At the same time, we experience reduced processing speed, accuracy, and productivity. Such trade-offs occur when we are multitasking and explain why light multitasking, such as listening to music while cooking, is easier and has fewer consequences than heavy multitasking, such as texting while driving.[26]

What has been less well understood are the mechanisms by which working memories are created and stored and what, if anything, we can do about the limitations. The last twenty years have seen considerable advances in this regard, as scientists have teased out the neurobiology and neurophysiology of working memory. We now can predict an individual's visual working memory capacity on the basis of neurophysiological signatures—specifically, EEG brain waves. Sharp spikes in EEG brain waves predict accurate working memory allocation, while a more diffuse or scattered pattern predicts a higher error rate. It is a wonderful example of how our language sometimes reflects our cognition: we describe focused attention as "sharp" and unfocused attention as "scattered," and this is analogous to what is seen in the brain.[27]

Researchers have used these neurophysiologic markers to study how distraction results in reduced working memory. Let's return for a moment to the metaphor of memory as physical storage—our cubbies represent our working memory capacity. Let's further stipulate that each of us possesses a limited number of cubbies. The question is whether individuals with lesser working memory capacity have fewer cubbies or whether they simply fill up their cubbies with junk rather than useful things. In other words, do people with less effective working memory suffer from a lack of memory capacity or from poor attention control and allocation, so that they do a poor job of separating relevant from irrelevant information?

It turns out that, at least in the case of visual working memory, the problem isn't a lack of memory capacity but rather poor attention control and our tendency to temporarily store junk in our brains. To figure this out, researchers gave a group of adults working memory tasks while also presenting them with a variety of distractors. Then the researchers isolated neural signatures of attention processing using EEG and eye scans. The results strongly suggest that working memory deficits are the result of filter failure rather than the lack of capacity to store information. When faced with distractions, adults with lower working memory capacity have more difficulty tuning out irrelevant stimuli and use up their memory capacity with irrelevant items. In contrast, individuals with higher memory capacity are much better at filtering out distractions and storing relevant information.[28]

With this in mind, the question becomes: Why? What separates high-capacity individuals from their low-capacity peers? What is going on in brains of those possessing better and worse attention control? To answer these questions, scientists have relied on neuroimaging, which can show us which parts of the brain appear to be involved in the filtering process and what else we might learn about attention control in distracting environments. You will not be surprised at this point to learn that the findings lead, once again, to the prefrontal cortex.

In a 2010 study, cognitive psychologist Andrew Leber conducted two neuroimaging analyses of healthy adults while they performed a visual working memory task. Each participant performed one task repeatedly, sometimes with a distraction present and sometimes without. In contrast to previous research that looked at individuals with low and high capacity for visual working memory, this study looked at the variability within an individual over the course of many trials of the same task. This is an important distinction: rather than compare individual performances, Leber looked at individual performance over time. The fMRI study confirmed that some people have greater visual working memory capacity than others, because some people are better at ignoring distractions. This suggests that the impediment to working memory is filter failure. However, Leber's study was able to show something else: even within individuals, there is variability. His study gives us the opportunity to see if there are neural correlates to this variability in working memory within ourselves and how working memory capacity might relate to attentional control in the brain.[29]

The neural correlates Leber found lie in an area of the prefrontal cortex called the middle prefrontal gyrus: the level of activity in the left middle prefrontal gyrus predicted task performance with a high degree of accuracy. Interestingly, the activity in the prefrontal cortex preceded the task. This suggests that when we are stressed out, overwhelmed, or engaged in too much multitasking, it gets harder to tune out distracting information and tune into and retain the information that matters most. The study shows that if the prefrontal cortex is not prepared before the activity then our performance will suffer.

We return to our metaphor of the prefrontal cortex as the conductor in our brain. The conductor provides resources and prepares the orchestra in advance of the symphony performance. But the conductor does not participate in the actual playing of the instruments. Likewise, the prefrontal cortex does not do all of our information processing and isn't responsible for carrying out actions, but the more prepared our prefrontal cortex is for the task, the better direction other areas of the brain receive and the better our performance. When our brain is in a healthy state, we are more equipped to handle whatever comes our way. The conductor adroitly commands some sections to play while others stay quiet. When our brain is overwhelmed or stressed, our ability to focus on relevant information decreases, and we are susceptible to distracting stimuli resulting in a form of filter failure.

This is yet another demonstration of the importance of a healthy prefrontal cortex in our increasingly distracted world. Linda Stone, a tech writer and former researcher and executive at Microsoft, describes this world as one of *continuous partial attention.* Continuous partial attention is a state of mind in which we are so anxious not to miss any of the overwhelming range of available stimuli that we are reluctant to give any one thing our full attention. As education and media scholar Ellen Rose puts it, we are "constantly surveying the infoscape—even, or perhaps especially, when we are supposed to be doing something else."[30]

As the online world provides more and more information—information that may be irrelevant to whatever task is at hand but designed to grab our attention and trigger our reward centers—we get too easily distracted in our offline world. We give in to every app ping and message ring and are more prone to different types of multitasking. The prefrontal cortex becomes stressed and tired. We lose the ability to conjure up sharp focus. Task completion suffers and memory suffers with it because the information we are attending to is the information that enters our working memory. Long-term memory and learning can suffer, too, because short-term memory is the precursor to long-term memory. In the end, we all suffer the consequences.

Media and Technology Multitasking Revisited

In 2009 Clifford Nass of Stanford University published a study whose results surprised him and his colleagues. "We all bet high multitaskers were going to be stars at something," Nass said in an interview. "We were absolutely shocked. We all lost our bets." It turns out the young-adult multitaskers he studied were terrible at nearly every aspect of multitasking. We now return to a major theme of the book: the consequences of media and technology multitasking, but this time with a view toward the adult brain.[31]

Nass's study is important in the multitasking literature for three reasons. First, it set a high standard for the rigorous measurement of attention and cognitive control in media multitasking research. Second, it was one of the first in a series of studies showing that performance consistently declines during multitasking. Third, unlike most research papers, where authors report findings that are consistent with their initial predictions, Nass made clear that his expectations were dead wrong.

Not that his expectations were outlandish. In fact, they reflected conventional wisdom at the time. As media and technology multitasking were on the rise, neuroscientists and cognition specialists assumed these activities would confer some benefit on those who engaged in them repeatedly. This was, as we saw, presumed of adolescents; why not adults as well? Years of multitasking *must* result in some improvement in performance in some domain—why else would so many people be doing it?

Yet Nass and his colleagues found that study participants classified as heavy media multitaskers were more susceptible to interference from irrelevant information. They also performed worse on a test of task-switching ability, suggesting weaker cognitive control relative to light media multitaskers. Of the heavy media multitaskers, Nass told an interviewer, "They're suckers for irrelevancy. Everything distracts them." While the research could not establish causation—that is, whether people who have poor cognitive control end up as heavy media multitaskers, or whether more multitasking leads to poor cognitive control—it offered a cautionary tale.[32]

And, in the case of excessive multitasking, questions of causality do not factor in to how we might solve the problem. If, at every stage of development, our brains have limited capacity and are affected by multitasking, with little or no upside, and multitasking results in a form of filter failure in response to information overload that significantly harms our cognitive abilities, then the obvious (if not easy) intervention is to reduce the amount of media and technology multitasking we engage in. If, on the other hand, some of us are wired to be more susceptible to filter failure and weak cognitive control—because, say, we struggle with ADHD, chronic stress, or anxiety disorders—and we cope through media and technology multitasking, the same recommendation would apply. Research consistently suggests that too much media multitasking is likely to make attentional issues worse at every life stage, so we need to do less of it. But instead of doing less media and technology multitasking, we are doing more than ever.

We saw in the case of children that media multitasking had the potential to influence the development of the prefrontal cortex—the last part of the brain to fully mature, the brain region most vulnerable to media and technology effects, and the region most important for our ability to fully function in the world independently as adults. But adult brains suffer as well. Whether we are trying to perform two tasks simultaneously or switching back and forth between tasks—say, completing a work project while checking social media or news

126 REWIRED

alerts—the costs of this type of mental juggling are high. Some experts estimate that even a few tenths of a second decrease in attention can add up as we move back and forth between tasks, with impaired mental shifts resulting in a 40 percent loss of productivity at work. Adults and kids struggle to keep up while multitasking for the same reasons. Neither immature nor mature brains can recruit additional cognitive resources in order to help perform two tasks at once; the prefrontal cortex inevitably ends up with a bottleneck. We all have our limits.[33]

IN BRIEF: COMPUTER VISION SYNDROME

The human eye and body evolved outdoors. The hunting and foraging that sustained early humans were made possible by reliable distance vision and constant movement. But modern life demands the opposite use case for our eyes and bodies. Rather than focus on a mixed contour of distant images, with an ever-changing focal point that requires near-constant large-field accommodations while we move about our environment, the use of screens large and small sets our focal point at a fixed, short distance for prolonged periods while we sit relatively still indoors. Simply put, we did not evolve for the digital age.

Sitting still for hours on end with one's head and neck in a relatively fixed position contributes to back and neck strain. The use of a computer keyboard, mouse, and trackpad, combined with poor ergonomics in some settings, contributes to wrist pain. However, while neck, back, and wrist pain are problems associated with prolonged computer use, by far the most common complaint surrounding such use is related to the eyes, giving rise to a new problem in the digital age known as computer vision syndrome. Symptoms include eye strain, eye pain, visual fatigue, and redness. In more severe cases, symptoms can include blurred vision, double vision, burning sensation in the eyes, dry or watery eyes, and eyelid tics.[34]

The number of people affected on a global scale by computer vision syndrome may well be staggeringly high. Estimates suggest that up to 90 percent of people who use a computer for three or more hours a day may suffer, a massive total considering that individuals across a growing number of professions and at every level of their careers do their jobs with computers and other screens. This is not a problem only of "knowledge workers" or high-income countries. Computer use features in professions of all kinds, massively altering work habits around the world.[35]

The direct causes of computer vision syndrome have not been firmly established, but a few factors are thought to be at play. Experts believe that staring at a fixed focal point for hours taxes the visual system in a number of ways, and that the nature of the image provided by the computer screen itself also contributes to the problem.

One issue is that screens do not offer the edge contrast that ink on paper does. This is a product of how screens work: screen images are built from pixels. At a typical focusing distance, digital screens appear to show smooth, continuous lines, curves, and fields of color, but the images we see are not in fact continuous. Rather they are built from tiny areas of illumination. Pixel-based displays offer far less contrast than ink on paper—that is, there is less contrast between the edge of, say, a text character on a screen and the white background projected around it than between a dark letter printed in ink and the typically off-white color of the surrounding paper. The lower contrast of computer images makes the edges of the characters less well defined, so that the visual system has to work harder to focus.

Low contrast and less-defined edges trigger three visual adaptations, which happen without our conscious awareness. First, the eyes relax a bit in response to the low contrast and therefore attempt to fixate on a point behind the screen. As a result, we are constantly and subtly refocusing on the surface of the screen. Second, in an attempt to sharpen focus on low-contrast text, we compensate with myriad small eye movements and minor near-field accommodations. Third, in a final attempt to increase focus, we squint. Squinting has been shown to reduce the number of blinks by up to 50 percent, contributing to dryer eyes. More demanding work with a higher cognitive load—for example, technical reading—further reduces blink rates, increasing eye strain. As we engage in these adaptations over the course of hours, days, weeks, and years, we fatigue our eye muscles, resulting in uncomfortable symptoms.[36]

What can we do about it? Some experts call for public education, government guidelines, additional training for eyecare specialists, workplace-safety regulations, more work breaks, and more research on the topic. But while we wait for governments and employers to catch up, we can all adopt some basic, commonsense recommendations.

First, consider the ergonomics of your workspace. The screen should be 20–25 inches away from your face to minimize strain, with the center of the screen 4–8 inches below eye level to minimize squinting and reduce neck strain. To compensate for the comparatively low contrast of computer text, keep the ambient room light lower than the brightness of the screen. This increases apparent contrast. Further, wearers of corrective lenses should be sure to have regular checkups rather than simply waiting until prescriptions no longer seem adequate. And all heavy screen users should be sure to take regular breaks— preferably outside, where the focal point is at a greater distance and we get exposure to sunlight. Some ophthalmologists suggest the "20–20–20" rule— every 20 minutes take at least a 20 second break and focus on something at least 20 feet away. Finally, use saline eyedrops and a humidifier to moisturize your eyes and your environment, and try consciously to increase your blink rate when doing more demanding computer tasks.

IN BRIEF: TECHNOLOGY IS A PAIN IN THE NECK

A few years ago, after a stressful period, I strained my lower back lifting my then-two-year-old son out of the car. I could barely move from the pain. Medication proved to be of little help, and hours of research on back pain yielded few ideas that worked. Eventually I decided to see a chiropractor. To my surprise, my chiropractor was concerned not only about my lower back, but also my neck.

I learned that years of studying, combined with endless hours of computer and smartphone use, had taken a physical toll on my neck. For the moment, the back pain seemed more severe, but stiffness and poor range of motion in my neck, alongside occasional painful spasms, were signs of worsening symptoms ahead. My chiropractor recommended exercises to increase strength and flexibility in my lower back and in my neck.

Neck pain is another growing epidemic. Musculoskeletal disorders of the neck and spine are on the rise due to poor posture while engaged in mobile computing and prolonged sitting in front of screens. As students, workers, and employers have increasingly recognized the problem, a thriving ergonomics industry has emerged, focusing us on desk height, chair comfort, and screen position. Stand-up desks and flexible screen armatures have grown in popularity, while office chairs have sprouted a vast array of knobs and paddles we can use to adjust our seats every which way to reduce the strain of hours of sitting.

But while there are endless ergonomic fixes for the office environment, almost no one has an answer for "iPosture" and "text neck," terms appearing in the medical literature with increasing frequency. A 2017 study of over 5,000 young adults in Sweden showed a significant relationship between frequency of text messaging and musculoskeletal symptoms including neck and arm pain. The results suggested that younger age of onset resulted in continued symptoms over a five-year period.[37]

Why? Imagine your head weighed fifty pounds. That would be a massive load on your neck, which is accustomed to holding up something more like twelve pounds, the average weight of an adult head. But this is what the neck experiences when you tilt your head down forty-five degrees to stare into your smartphone or tablet. Even if you tend to stick to a thirty-degree angle, you will still be putting forty pounds of pressure on your neck.[38]

This extraordinary weight, experienced for long periods of time, fatigues the soft tissues, ligaments, and muscles in the neck and shoulders, resulting in inflammation. This in turn results in pain, soreness, and eventually can lead to pinched nerves, disc herniation, muscle stiffness, decreased range of motion, and other serious consequences for the neck and spine. "It's an epidemic, or at least, it's very common," orthopedic surgeon Kenneth Hansraj told the *Washington Post*.[39]

Poor posture and spinal problems have been linked to other serious issues, including headaches and heart disease. Neck pain is also associated with mental health problems. One study in Germany found that, among patients in a primary care practice, there was a strong association between neck pain and the clinical levels of depression or anxiety, independent of other factors.[40]

What should we do about it? First, when using your phone, try holding it up to keep your head level rather than bending at the neck. This relieves neck pressure and keeps your eyes more relaxed, which can reduce eye strain. Second, do daily exercises such as stretching your neck as part of your regular breaks from screens. Start by moving your head forward with your chin in your chest and then backward with your eyes toward the ceiling a few times. Then press against your forehead with your hands as you push your head forward. The forehead pressure creates resistance that stretches and relaxes the neck muscles. Then do the same course of exercises while pushing your head backward and side to side. Rotating the neck is also useful. In addition to strengthening neck muscles, these motions keep muscles flexible, which reduces inflammation. Third, stand in a doorway with your arms outstretched while pushing your chest forward and keeping your head up. This puts your neck in alignment with your spine, adjusting your posture and helps keep shoulder muscles limber.

Finally, we all need to be more aware of where our heads are from time to time. That's good advice literally and metaphorically. Keeping your head level on your shoulders is good for neck health. And being attuned to how you feel goes a long way, too. Don't ignore signs of strain. Seek medical advice. Unlearning the habit of bad posture is easy, as long as we pay attention to the signals our bodies are sending us.

7

IMPACT ON MENTAL HEALTH

You have heard stories, perhaps even been a victim. A couple, previously in love, has a particularly bitter breakup. Feeling scorned and jilted, the ex-boyfriend posts nude or seminude photos of his former girlfriend on social media, along with unflattering remarks. She is helpless, horrified, and becomes depressed. She is a victim of revenge porn, an increasingly common form of cyberbullying unimaginable in the days before smartphone cameras and online social networks that thrive off photos and videos of people in compromising positions.

Reliable statistics on harassment online—which includes denigration, exclusion, trolling, and other tactics—are difficult to find. But studies estimate that between 20 and 40 percent of children and adolescents have been victims of some form of cyberbullying. Women, LGBTQ people, and obese people are common targets. And while cyberbullying appears to be less prevalent with older age, adults are at times victims, whether of direct harassment or past trauma: vicious online treatment in childhood and adolescence can manifest in adults as post-traumatic stress disorder, depression, and suicidality.[1]

Experts debate the exact definition and the full scope of cyberbullying, but at its core, it is a failure of empathy. Empathy refers to our ability to read or understand accurately the mentalizations and emotional world of another person. It is not the same as sympathy, an older term that today is more aligned with feelings of sorrow when someone else is experiencing misfortune. Sympathy is self-oriented; its perspective is one's own. For example, one might feel sadness after learning that a friend broke her arm. Empathy, by contrast, is other-oriented; it takes the perspective of another person. Rather than feeling sorrow for the friend who broke her arm, we feel her pain and imagine what it might be like for her, which promotes feelings of concern.

Most social psychologists view empathy as a key human trait that is required for cooperation, altruism, compassion, and other prosocial skills. It is the basis of strong relationships and essential to our survival. As social creatures, humans rely on empathy to form the bonds that allow infant brains to develop and that facilitate both independence and community later in life. The lack of empathy is a hallmark of antisocial behavior, psychopathy, and predatory violence. Too little empathy often leads to dehumanization and callous disregard for the well-being of others, contributing to online harassment and other types of aggression, as well as societal problems like political divisiveness.[2]

Early signs of empathic concern have been documented in humans as young as six to eight months old. But this critical prosocial skill develops well into adulthood, which is not surprising given that empathic abilities involve the prefrontal cortex. One area of the prefrontal cortex, the ventral medial prefrontal cortex, has been associated with both empathy and the development of moral judgment. Could excess media technology use disrupt areas of the prefrontal cortex during childhood and early adult development to the point of reducing empathy and moral judgment?[3]

Sara Konrath is a social psychologist who specializes in understanding empathic abilities. In 2011 she and several colleagues published a widely cited study that looked at empathic traits in college students over a forty-year period. The research synthesized and analyzed findings from more than seventy studies of 13,000 young adults in the United States from the 1970s into the 2000s. Her team used a well-validated multidimensional self-report scale that measures two aspects of empathy: empathic concern and perspective-taking. To evaluate empathic concern, participants are asked whether they agree with the statement, "I often have tender, concerned feelings for people who are less fortunate than me." People who score high on empathic concern are more likely to engage in prosocial behaviors such as volunteering their time, giving money to a homeless person, and returning incorrect change. Perspective-taking—the ability to put oneself in another's shoes, more or less—is also related to prosocial outcomes. Low scores on perspective-taking have been correlated with a propensity toward violence and criminal offenses.[4]

Konrath and her colleagues found a substantial drop in empathic concern (48 percent decrease) and perspective-taking (34 percent decrease) over the forty-year period studied. Interestingly, the decreases were sharpest between

2000 and 2009. The study authors speculate that "one likely contributor to declining empathy is the rising prominence of personal technology and media use in everyday life." On this view, the decrease in scores, while relatively small, reflects a broad shift toward a more self-oriented and shallow kind of communication promoted by our favorite digital tools. It is no coincidence that in the final decade of the analysis, when empathy scores dropped the most, we saw the launch of the iPhone (2007) and social platforms including Friendster (2002), MySpace (2003), Facebook (2004), YouTube (2005), and Twitter (2006).

The idea here is that we are all spending less time in synchronous, one-on-one, face-to-face communication, with its rich verbal and nonverbal signals that contribute to empathy. Now our communications are shorter, asynchronous, one-to-many, and mediated by instant messaging and social media. We increasingly substitute emojis for emotions. This change from warm and intimate to cold and distant communication runs the risk of detaching us from the deeper connections we have evolved to depend on, potentially contributing to cyberbullying and other bad behavior online, where consequences are often minimal.

But this is just speculation. What does the evidence say? Is our use of mobile media, communications, and information technology interfering with face-to-face social experience and thereby harming our capacity for empathy?

Alone Together

You are sitting with a friend or loved one, engaged in face-to-face conversation on a topic of some importance. Suddenly your smartphone makes a noise or vibrates in your pocket. It is impossible to ignore. Even if you don't look at your phone—even if it occurs to you that doing so could insult the other person—the flow of conversation is disrupted, and you miss a beat while contemplating whether the notification you just received was more important than your companion's feelings. Does this type of micro-disruption matter?

Two studies strongly suggest that it does—the mere presence of a smartphone can negatively affect one-on-one, face-to-face social interactions. One of the studies, published in 2014, concerned what researchers called the "iPhone effect." The scientists recruited a hundred groups of social pairs—friends who knew each other before the study—in the Washington, DC, area. The friendship pairs were recruited on their way into a coffee shop and asked to sit across a table from each other and chat.

The researchers randomly assigned each pair to converse about one of two topics for ten minutes. One topic was superficial; the study participants were prompted to "discuss your thoughts and feelings about plastic holiday trees." The other topic was more emotionally laden; participants were asked to "discuss the most meaningful events of the past year." As the two friends talked, a trained observer rated their nonverbal interactions and communications. The observer also noted the presence of smartphones, where each of the participants put their phones, and whether they used them while conversing. Finally, the observers rated the conversations using a scale of empathic concern and a scale of psychological closeness, which were used to evaluate the quality of the interaction.[5]

Twenty-nine of the hundred pairs conducted their conversation while some mobile communications technology was visible. Controlling for other variables, the authors found that the visible presence of technology had a significant correlation with lower empathic-concern and psychological-closeness scores. Importantly, the device did not have to buzz, beep, ring, or flash. Its mere presence in hand or on the table was sufficient to produce a measurable drop in the quality of the interaction.

The researchers speculate that this drop in quality was related to a type of distraction sometimes referred to as divided attention: a smartphone represents access to a wider social network, email, news, entertainment, work, and endless information, triggering a "constant urge" to check for updates and communications, even when we are face-to-face with a friend. The phone directs our thoughts to other worlds and other people—people other than those right in front of us.

Once again, some caution must be taken as we interpret these results. We cannot infer causation: it is possible that people who are more likely to place their mobile media technology in view or hold it during a social interaction are more distractible and less empathic in general. Fortunately there are other data that, when placed alongside this study, can help us understand whether the mere presence of smartphones has a causal effect on the quality of in-person interactions.

The research on friends in coffee shops was inspired in part by prior research with strangers in a more controlled situation. This study, conducted just a couple of years earlier in the United Kingdom, involved participants who were randomly assigned to have a ten-minute conversation with a stranger in one of two conditions in a lab environment. In the first condition, a nondescript mobile phone was placed on a table between the two chairs where the participants sat for their conversation. In the second condition, the mobile phone was replaced by a blank notebook. As with the coffee shop study, the discussion was meant to foster some level of intimacy and social connection, so participants were told to discuss an "interesting" event from the past month. The researchers found that when the phone was present, conversation participants achieved a reduced closeness score.[6]

So the visible presence of a smartphone appears to affect one-on-one, face-to-face conversations across studies testing a range of conditions. There appears to be a form of technology multitasking at work here, whereby micro-disruptions divide our attention between the real world and digital world so that we cannot fully focus on what is happening immediately before us. Perhaps we experience filter failure, so that we are paying attention to the wrong stimuli and therefore miss spoken words, facial expressions, or other nonverbal gestures that contribute to the depth of social connections. And as a result, we are less empathic.

Both of the research groups involved in these studies refer to the work of Sherry Turkle, an MIT professor who has been studying and theorizing about our relationship with technology and the internet for nearly thirty years. A leader in research at the intersection of science, technology, and society, Turkle has written several acclaimed books. She has also collected numerous anecdotes illustrating how the nature of conversation is changing as a direct result of our relationship with technology. These anecdotes are no substitute for detailed, well-designed studies like those described above, nor for a big meta-analysis of the variety that Konrath and her colleagues undertook. But, alongside rigorous research, the stories are telling.

Witness some examples from Turkle's work. She recounts a middle school administrator aghast at the lack of empathy he observes on the playground at recess: "Twelve-year-olds play on the playground like 8-year-olds. The way they exclude one another is the way 8-year-olds would play. They don't seem able to put themselves in the place of other children." It is as though the twelve-year-olds have regressed developmentally. Another teacher notes that her students, all of whom are using media technology more frequently, sit in the dining halls and look at their smartphones instead of each other. "When they share things together," she adds, "what they are sharing is what is on their phones."[7]

Other examples abound. Not long ago, I was at a restaurant in Boston waiting for a friend. I had been at the bar alone for some time when four young women came in for dinner. They chatted energetically, exchanging glances, gestures, and stories from their lives. Clearly these were people who knew each other well. Within minutes of sitting down, all four of them had their phones out and were busy typing, lost in their online worlds and ignoring the real world in front of them. The social bond that was so obviously warm as they walked in the door was suddenly cold and broken. Their expressions turned blank. Eye contact decreased. Energy levels dropped. Conversation halted. Their smartphones were disrupting the gathering at a meaningful moment—another case of what Turkle calls, in her book by the same title, being "alone together."

Social Brain Drain

We see from the research how the mere presence of a mobile phone creates a type of split attention and can affect the quality of social interactions. Amazingly, in the laboratory experiment, the phone did not even belong to either participant. It was simply a mobile phone on a table. Neither of the participants had reason to worry that they were missing a message on the phone. The presence of *any smartphone* can undermine our attention to each other. This is a new type of social brain drain.

Indeed, it is not only social interactions that are subject to this brain drain. Other types of information processing can also suffer from the presence of a nearby smartphone. This was the finding of a remarkable study by researchers at the University of Texas, Austin. The researchers measured the cognitive performance of over 500 undergraduates as they tackled difficult math problems while simultaneously trying to remember a sequence of random letters. The students were assigned to one of three conditions. In the first condition, they were told to leave their smartphones with their belongings outside the testing room. In the second condition, they were told they could keep their smartphones with them but out of sight, in their jackets or pockets or under the chair they sat in. In the third condition, the students were told to keep their phones on the desk in front of them but facedown and with the ringer and vibration functions turned off to minimize distractions.[8]

Here, too, the mere presence of a smartphone appeared to impair information processing: greater proximity and visibility were associated with worse outcomes. The students who had the smartphone on their desks performed worst, followed by those who kept their phones nearby but unexposed. The students who left their phones outside the room performed best on the test.

How is it possible that the mere presence of a smartphone we are not using can alter our social interactions and our ability to process information? The answer lies in the phenomena of salience and relevance. Salience and relevance influence how the brain allocates its limited resources.

To grasp the distinction between salience and relevance, let's return to the visual system. One of the leading theories of how visual processing works suggests that we continuously create and update a mental map of our visual environment, and we do so using a network of specialized neurons. Some of these specialized neurons help the eyes move in order to scan our environment. Other neurons help process the physical appearance of an object. And a third group of neurons responds to both the appearance of an object and its movement. Collectively, these specialized neurons evaluate the features of the objects in our environment and help us pick out the most salient and relevant objects.

Salience refers to the physical qualities that make an object noticeable—the features that make it stick out from the crowd. These features include distinctive color, orientation, size, shape, and motion. Salience, importantly, is not inherent in objects. Rather, it arises from the contrast between an object and its surroundings. Fast motion is not inherently salient; fast motion is salient against a static or slow-moving background. Laboratory studies of salience may ask participants to find the red dot among a group of blue dots, for example.

To say that an object in the visual field is salient is not to say that we have decided that it is important, only that it has caught our eye. We in fact make no conscious decisions with respect to salience, because salience is determined without our conscious awareness. Salience is a matter of bottom-up processing; it occurs automatically, relying on specialized neurons and deep brain structures that require very little effort from the prefrontal cortex. The more salient the features of an object, the more these specialized visual-processing neurons respond. The more neurons that respond, the more likely they are to contribute to a summing-up process in the brain. If a certain threshold of neuronal activity is hit, we direct our attention toward the object without hesitation. This is known as automatic attention.[9]

Some salient objects are also relevant and may indeed be salient because they are relevant. These are objects that we do not merely notice but prioritize. Relevant objects in our environment—and there are lots of them—are familiar from prior experience and are likely to provoke emotional associations. Objects gain relevance and therefore attentional priority because we encounter them repeatedly, because they are useful to us, and because they present to us some emotional reward. Relevant objects may advance goals such as avoiding harm, maintaining health, and achieving reproductive success. They may serve some need or they may demand some action in response to their presence. Such objects may be physically distinctive, but they need not be. They stand out because our experience with them makes them emotionally relevant and therefore their salience increases.

An illustrative example of something whose salience follows its relevance is one's name. Objectively speaking, the sound of a name is not salient. However, imagine you are at a crowded cocktail party, a noisy, distracting environment where many conversations are going on at once. Your brain will not attend to much of what is being said in these assorted conversations beyond your own, including the various names that the partygoers mention. But your brain will definitely take note if you hear your own name spoken across the room, and you will shift your attention to its source. This is an example of automatic attention and what's called the cocktail-party effect: the ability to focus auditory attention on one stimulus among many in a crowded backdrop. Neuroimaging studies suggest this phenomenon relies on a brain network that includes the prefrontal cortex and auditory processing areas. Because our names are relevant to us, they command our attention instantly.[10]

What appears to be going on with our smartphones is that their relevance makes them salient, and as a result they command our automatic attention. Endless repeated use and resulting emotional rewards have turned otherwise-nondescript slabs of metal and glass into automatic attention-grabbers. Automatic attention is wired into our brains because it aids us in allocating limited cognitive resources; we don't have to consciously direct our attention or think about specific goals in doing so. But when automatic attention falls on objects that don't serve an immediate goal, we become distracted from what matters.[11]

Marketers understand this well, which is why they invented "bells and whistles." It's not by accident that this phrase refers to salient phenomena: loud, high-pitched noises that stand out from the background. Buzzers and blinking lights automatically grab our attention, even if they serve no purpose. The same is true of the clickbait that sits next to news headlines online. Clickbait is irrelevant to the task of searching for information or reading the news, but if the clickbait is well crafted, it will nonetheless be salient, attracting our attention and distracting us from the initial task. Facebook newsfeeds are masterpieces of combined relevance and salience. Facebook's algorithms ensure that our feeds are full of relevant posts—material from friends and other sources we know well, interspersed with content that dovetails with our genuine interests. At the same time, Facebook's engineers have built in features like notifications and autoplay videos, which are designed to stand out visually, achieving salience.

Facebook's combination of salient and relevant features makes it a potent engine of distraction; as we saw in our examinations of multitasking, social media is especially taxing on our cognitive faculties, causing task performance to suffer greatly. But we don't even need to be using social media to realize the cognitive consequences that lead to micro-disruptions in our connection with others and with our work. The mere presence of a smartphone in our visual field implies access to social media and all the other tools that smartphones offer—tools whose rewards have trained us at a nonconscious level in new habits of automatic attention.

Habits, however, have consequences. And the trouble is that many of our habits around mobile media, communications, and information technology are maladaptive, dividing our attention in ways that hamper our social skills and cognitive capacity, contributing to a growing social brain drain. No doubt smartphones and the tools they provide have many benefits. That is why phones are both relevant and salient. But the benefits come at significant cost.

The New Age of Anxiety

You're sitting alone after a stressful few days when suddenly you notice your heart is racing and your breath is getting short. You try to stop the maddening rush of thoughts. Your chest is tight, maybe even painful, and now your palms are sweating as you wonder, "Am I having a heart attack?" After a trip to the emergency room, the symptoms dissipate, and the doctor reassures you that your heart is fine—it was a panic attack.

Anxiety is clinically defined as exaggerated worry or fear over matters that would not be a cause for serious concern for an otherwise-healthy person. The signals that trigger anxiety start in the emotion centers of the brain and are misinterpreted by the prefrontal cortex as a threat. Those who experience clinical anxiety are unable to tamp down or reinterpret these signals correctly. While all of us experience a certain degree of anxiety, in the pathological state, where the neural circuitry is clearly disordered, anxiety can be disabling.

Exaggerated worry is key here: the level of fear in cases of clinical anxiety is overblown, but there is usually some realistic basis for worry. In the case of a panic attack, the sufferer might fear having a heart attack, which often involves similar symptoms such as a racing heart, chest tightness, and sweaty palms. Some people experience extreme anxiety when driving over a bridge or getting into an elevator—perfectly normal activities, albeit ones that involve some risk. Some of my patients who have clinical anxiety in the form of obsessive-compulsive disorder can't help worrying about whether they left the stove on or locked the front door, even after confirming multiple times that all is in order. The root of these patients' fears is reasonable; no one wants their house to catch fire or be robbed. What isn't reasonable is their struggle to reassure themselves despite knowing that they have taken adequate precautions.

Today we are living in a new age of anxiety, thanks in part to the digital revolution. The digital revolution directly alters our internal lives—our relationship with information, our identities, our psychology, and ultimately our brains. New methods of creating, storing, managing, and distributing information are not just changing how we work, live, and play. They also influence our thoughts and emotions. Our prefrontal cortex has a hard time keeping up with these changes, and anxiety increases.

A 2018 national poll of adults by the APA showed the highest levels of anxiety in years. By generation, Millennials were more anxious than Gen X or Baby Boomers, but Baby Boomers, the oldest generation polled, showed the greatest year-over-year increase in anxiety. "That increase in stress and anxiety can significantly impact many aspects of people's lives including their mental health," said APA president Anita Everett, discussing the poll results. According to data from the National Institute of Mental Health, anxiety is running high among US teens, too, with one in four boys and a whopping 38 percent of girls diagnosed with an anxiety disorder. Among adults, one in three Americans will be affected by an anxiety disorder during their lifetime.[12]

There are several major sources of anxiety in the United States today, including health, safety, and finances. These anxieties are heightened by lack of relaxation that comes with an always-on culture, which is in many ways a product of our broken tech-life balance. As we have seen, digital distractions and the demands of a tech-driven 24/7 work life have a tendency to keep us up at night and to invade social and family time. It is increasingly impossible to break away from work even while on vacation. Indeed, this is among the reasons Americans take so little time off: in 2015 Americans left 658 million vacation days unused. Why not take time off? Financial pressure is a serious issue for some, but even Americans with paid time off leave vacation days on the table. One cause is the expectations fostered by technology—the daunting pressure to check work messages while on vacation or else face a mountain of emails upon returning to the office, poisoning the experience of time off. Either vacation isn't really vacation, or it is fraught with anxiety.[13]

It is no wonder that, as a 2017 *New York Times* headline puts it, "Prozac Nation Is Now the United States of Xanax." The *Times* notes a stunning profusion of books, TV shows, online commentary, and even Broadway shows reflecting a nation that is not only depressed but also filled with anxiety. Cultural divisiveness, political polarization, climate change, economic uncertainty, information overload, and loss of privacy all fuel worry about the future. Push notifications. Apocalyptic headlines. Rancorous tweets. The article quotes one social media consultant, a profession that did not even exist ten years ago, stating, "If you're a human being living in 2017 and you're not anxious, there is something wrong with you." The same might be said today.[14]

The Social Media Paradox

Jennifer Garam is a Brooklyn-based blogger and self-professed coffee-drinking, yoga mat–toting, bookstore-browsing writer, teacher, and speaker. She is also incredibly honest about her feelings around social media. "When my self-esteem is shaky, which it often is, I have to be careful around social media," she writes. "On Facebook and Twitter, everything is always wonderful for everyone and all their lives are amazing." She admits to struggling with depression, and social media makes the problem worse.

Garam bemoans the curated lives depicted on social media—the continuous stream of baby bumps, beautiful children, and statements of eternal love on anniversaries; the photos of joyous families on fabulous trips and the announcements celebrating job promotions. To her it seems that everyone is having the proverbial best time ever. "How are people taking all these interesting pictures from fascinating places all the time?" she wonders, her exasperation mounting.

"Where are they? Doesn't anyone else have to work? *WHY IS EVERYONE'S LIFE SO MUCH BETTER THAN MINE?!?!?!*"[15]

So far, I have discussed how smartphones, always-on work cultures, media and technology multitasking, and online gaming and internet pornography addictions can imperil adults' productivity and cognitive capacity. Now I turn to the relationship between adult mental health and social media specifically. Although adults have the benefit of a mature prefrontal cortex, ample data show that we too can suffer from misuse and overuse of social media, including exposure to the kinds of social comparisons that vex Garam. If we are not going to cut social media out of lives, then we need to be mindful of both the quantity and quality of our use to limit its impact on our moods and mindsets.

At the heart of the matter lies a deep irony: a technology whose primary purpose is to connect people socially and foster communication—two foundations of mental health—can result in decreased socialization and increased mental health problems. How can this be? This question actually predates smartphones and social media, having emerged in the late 1990s, as people were first going online in droves. Much of that use was for email, blogging, and posting comments in chat rooms and forums. At least in theory, these tools were meant to connect people over long distances more efficiently, enabling them to share ideas, advice, news, gossip—anything that could be transmitted across a relatively slow data connection.

Researchers immediately saw opportunities to investigate what life was like before and after individuals had begun using the web, in hopes of answering enduring questions about the relationships among new technologies, socialization, and basic human values. One early study, from 1998, found that greater use of the internet was associated with declines in communications with family members and smaller social circles in real life. The authors speculated that time spent connecting online was displacing time for socializing offline. The study also found an association between time spent online and an increase in depression and loneliness, suggesting that online socializing was a poor substitute for the offline alternative. In response to these surprising results, they coined a term: the internet paradox.[16]

More detailed follow-up research a few years later suggested a more nuanced conclusion, however. Consistent with the "rich get richer" idea, internet use was associated with better outcomes for extroverts and those with more social support but worse for introverts and those with less support. This suggests that our predispositions; the particulars of our personalities, families, and community lives; socioeconomics; and other factors help to determine what constitutes healthy online experiences—both how much is good for a given person and how that time should be spent.[17]

Over time the internet paradox became the social media paradox, as researchers' focus shifted to whether use of social media, as opposed to internet use in general, makes us less social and affects our well-being and mental health. The research findings tend to be more equivocal than those of 1998. This makes sense, given that there are so many more uses of the internet and mobile technologies and the user base is much larger and less self-selecting. So it should not surprise us that when it comes to adults, a clear consensus on the impact of social media remains elusive. But even without a consensus, we can take important lessons about how best to manage our online social networks and ensure that they don't create problems for our mental health.

It is essential that we take this problem seriously. As discussed below, people with a genetic predisposition or other risk factors for depression—for example, those with strong family history, past trauma, or problems with substance abuse—need to take special care in how they use social media platforms that can amplify their symptoms. The increasing assault of mobile media, communications, and information technology, combined with the stress of our 24/7 lifestyles, is undermining our prefrontal cortices and their ability to regulate our emotion centers, manage negative thoughts, interpret threats accurately, and fend off a rising tide of online social comparisons. This may be putting more people than ever at an increased risk for depression, contributing to rising rates of the disorder and its most serious consequences.

Social Media and Well-Being

One study worth examining in detail comes from Holly Shakya, a public health specialist, and Nicholas Christakis, a physician and sociologist who has been studying the impact of social networks on behavior for many years. (We met his brother Dmitri, a pediatrician, earlier in the book.) Shakya and Christakis set out to add to the literature on the impact of Facebook on mental health.[18]

They acknowledge that, so far, research concerning the impact of social media on adult well-being has been mixed. On the negative side, they cite studies showing that social media use is associated with increased risk of clinical depression and anxiety; displacement of intimate face-to-face interactions; reduced investment in other meaningful activities; increased screen time, resulting in more sedentary behaviors; more media multitasking; and the erosion of self-esteem. On the plus side, social media, and Facebook in particular, have sometimes been shown to enhance our moods and social relationships. One study of German and American Facebook users looked at participants in their homes and in the lab and found that positive emotions were more prevalent than negative emotions while browsing Facebook and that the closer the social bond between the user and the person whose posts they were viewing, the more likely the user was to have feelings of happiness. And plenty of studies have found no relationship between social media use and depression or well-being.[19]

However, there are a number of problems with these studies. First, many of them are cross-sectional, meaning that they analyze use at a single time point, so the impact of use over time is not factored into the results. Second, most studies rely on surveys of users to measure social media use, which often produces biased results. In these studies, a more detailed look at what people were actually doing on social media is lacking. Third, as with many of the studies I've reviewed, causality is a concern. Are those who are more prone to depression more likely to use social media, and is that accounting for some of the results?

In their 2017 study, Shakya and Christakis did something unusual in the research on social media and well-being: a multiphase longitudinal study that measured actual use. They looked not at one group at one point in time but rather at three successive waves of people over the course of three years. The study involved more than 8,000 adult participants, who agreed to use a tool that tracked their actual Facebook usage—for instance, how many likes their posts received, number of links clicked, number of status updates per day—and regularly fill out a self-report scale of mental well-being. An additional feature of the study was the collection of information about the participants' real-world social networks. Participants were asked to approximate how many real-world friends they had, state how close they felt toward those friends, and estimate the amount of time they spent with friends offline.

While the strength of real-world social networks was positively associated with overall well-being, the amount of time spent using Facebook and interacting with a virtual social network was negatively associated with overall well-being. These findings stood up across all three waves of analysis. Interestingly, while the effects were relatively small, the negative impact of Facebook use was higher than the positive impact of real-world social networks. This means that the negative impact of heavy Facebook use over time can outweigh the positive protective effect of real-world social networks.

Two important conclusions follow. First, because it tracked the same people over time and included real-world social networks, this large study gives us a glimpse into whether people who are already struggling are more likely to turn to Facebook for solace. The study suggests that this is not the case. Of course, some struggling people use Facebook as a coping mechanism, but that does not explain the observed impact of Facebook use on well-being. By controlling for initial well-being, the researchers found that using Facebook was associated with an increased likelihood of diminished future well-being across all users studied. This means that those who had low initial well-being were not more likely to use Facebook and have their well-being further diminished. Second, consistent with the original internet paradox findings, Facebook use does not, overall, promote well-being.

This impressive study tells us a lot about what Facebook means for general well-being: not much. A small negative impact is discernible, while emotional benefits are nonexistent at the population level, if not in every individual case. But there is more to learn. After all, measures of well-being are not used in mental health, so the study does not address Facebook's impact on clinical depression and anxiety. And what about other platforms? Facebook may be the biggest, but it is one of many that count hundreds of millions of users.

Another study tried to get at these questions by testing the relationship among depression, anxiety, and multiple online platforms (Facebook, You-Tube, Twitter, Google Plus, Instagram, Snapchat, Reddit, Tumblr, Pinterest, and LinkedIn). Researchers looked at 1,700 young-adult (ages 18–32) users of social media and found that, while overall use was not correlated with mental health outcomes, the number of different social media platforms used was strongly predictive of depression and anxiety symptoms. Study participants who used seven or more platforms were three times more likely to develop depression or anxiety than were participants who used two or fewer platforms (the study controlled for other factors such as income, race, gender, relationship status, and education). This suggests that extreme social media use is a risk factor for mental health problems or, at a minimum, a warning sign that such problems are likely to arise. The authors also speculate that people who use many social media platforms are more likely to engage in media multitasking, the negative effects of which on attention and productivity may also contribute to depression and anxiety.[20]

Because this is not a longitudinal study carried out over time, we can't rule out the possibility that people who are prone to depression and anxiety are more likely to use multiple social media platforms. Do people with high levels of anxiety or depression use social media more, or does more use cause people to be anxious and depressed? A deeper dive is needed in order to sort out causality.

How Social Media Reinforces Depression

Joanne Davila is a professor of clinical psychology and a relationship expert. Her 2015 TED Talk, which has over a million views, describes the skills needed for a healthy romantic relationship. "We may know what a healthy relationship looks like," she says, "but most people have no idea how to get one and no one teaches us how to do so; and this is a problem." She describes several key social skills that healthy couples have and how these skills positively affect romantic and other relationships. Clearly she has given a lot of thought to the power of relationships and wants people to have the skills needed to achieve success in them.[21]

In 2012 Davila explored a different type of relationship: the relationship between digital social connections and clinical depression. She and her colleagues were concerned that many studies in the field were too narrowly focused on social media use in isolation or too broadly focused on internet use generally. Most studies excluded text messaging, an important communication tool. Most studies also did not address individual mental health characteristics that might put some people at a higher risk for negative consequences. She and her colleagues undertook two studies aimed at overcoming these hurdles. Together, the studies help to untangle the social media paradox. The research involved 600 young adults: one study was cross-sectional, a snapshot in time, and the other was longitudinal, unfolding over the course of several weeks.

Davila found that, rather than the quantity of online social connections, it is the quality that drives the association between social media use and depressive symptoms. In both studies, participants who reported fewer positive and more negative interactions during digital social activities—including text messaging, instant messaging, and social media use—also reported greater symptoms of depression. If this seems intuitive, it should. Lower-quality real-world interactions are also depressing.

The second study confirmed this, as the researchers compared online interactions to real-world interactions with friends and romantic partners. The same association arose: those who reported fewer positive and more negative inter-actions offline were more likely to report more depressive symptoms. A good deal of other research also supports this conclusion: problems with friends and romantic partners, such as feeling excluded, conflicted, or rejected, can con-tribute to depression. Davila and her colleagues conclude that digital social interactions may be another venue in which problematic relationships mani-fest themselves and affect mental health.[22]

An interesting aspect of Davila's study is the inclusion of a rumination scale. Rumination here refers to the tendency of depressed people to think about their depression—a lot. There is substantial evidence that supports an association be-tween negative ruminations, carried out either alone or with a friend, and de-pressive symptoms, so it made sense to consider whether the degree to which people ruminate is related to social networking online. The research team used a self-report questionnaire that assesses how frequently an individual engages in a variety of thoughts, feelings, and actions when they experience a depressed mood. The idea is to quantify how much respondents ruminate: to what extent do they analyze recent events in an effort to understand why they are depressed, and how much time do they spend thinking about how sad they feel? The study also included a co-rumination scale to assess the tendency to share sad feel-ings with friends.

There were three results to consider. First, consistent with numerous past studies, higher scores on both the rumination and co-rumination scales were independently associated with higher amounts of depressive symptoms. This is not surprising, as one of the hallmarks of depression is alterations in our thinking, including negative ruminations. Second, the researchers found that people who engaged in more co-ruminations also engaged in more online so-cial interactions. This suggests that social media is an outlet for sharing nega-tive thinking, possibly contributing to sustained or increased symptoms of depression.[23]

Finally, people who had more negative ruminations tended to have more negative online social interactions and were more likely to feel depressed after using social media. The authors conclude, "Taken together, these find-ings indicate that the tendency to ruminate, either alone or with friends, gets played out in the context of social interactions that do not involve face-to-face or even verbal interaction." Instead rumination plays out through social media channels and mobile technology–enabled communications such as instant messaging.[24]

This pair of studies suggests that, for those at risk of depression, social media can sustain a downward spiral. Online social networks facilitate excessive and repeated focus on problems, which in turn contributes to depressive symptoms in those who are vulnerable. We all need to be mindful not to fall into holes like this, but young people and women are especially at risk because these groups are more prone to negative ruminations.[25]

Executive function is also implicated here. A recent review of studies involving over 3,000 participants showed a significant relationship between excessive ruminations and poor executive function. In particular, those given to excessive ruminations were weaker in two areas of executive function: inhibition—the ability to filter out distractions—and set-shifting, which involves changing from one task to another quickly and easily. This suggests that habits like heavy smartphone use and multitasking, which hamper development and functioning of the prefrontal cortex and therefore executive function, put us at increased risk for negative ruminations and depression. This is true for all of us, at any age.[26]

Negative ruminations are both a sign of depression and a source of it, and social media amplifies the depressive effects of ruminations by contributing to the negative thoughts that depressed people can't shake. To better understand how this happens, let's take a closer look at how depression works in the brain.

For years, researchers have theorized that one of the main cognitive components of clinical depression is a strong tendency toward pessimistic interpretations of innocuous information, particularly if that information is of high personal relevance. Neuroimaging research has tied this relentless form of negative thinking to the amygdala, a subcortical brain region with deep connections to the prefrontal cortex. The amygdala plays an important role in emotion generation; for our purposes, think of it as a relevance monitor. When we are depressed, the amygdala becomes overly sensitive to threatening stimuli, making negative experiences seem excessively relevant.[27]

While the hyper-sensitive amygdala overreacts to both perceived and real threats, at the same time the prefrontal cortex is less active—this, too, is a feature of depression. The depressed prefrontal cortex is less able to reframe and regulate the heightened threat response. This explains, in part, why psychotherapists focus on reframing bad thoughts and why this sort of intervention has been shown to have clinical value. In essence, the therapist is trying to engage the depressed patient's prefrontal cortex, encouraging it to frame events in more positive light.[28]

What we have here is another imbalance in the brain: increased activity in the amygdala producing negative thoughts and feelings, and reduced capacity of the prefrontal cortex to counter these negative thoughts and feelings. Instead of filtering out possibly negative stimuli, we take them in, overinterpret their importance, and then reflect on them repeatedly, resulting in a vicious cycle of negative ruminations. It is as though the conductor is too fatigued to keep a key part of the orchestra playing the right music on time.

Social media has been shown to reinforce this imbalance. A 2016 study confirms that negative online interactions can significantly affect the incidence of clinical depression. The researchers surveyed young adults and found that bad experiences on Facebook were an independent risk factor for depression. Unwanted contact or misunderstandings in online conversations were associated with a greater than twofold increase in depression risk, and cyberbullying and other "mean" posts, including personal attacks, were associated with a greater than threefold increase. The researchers also found that users experiencing a larger number of negative interactions online were at higher risk for developing depression.[29]

Given that social media is a known mechanism for exacerbating symptoms of depression, it is fair to speculate that this aspect of our digital lives is one factor underlying increasing rates of the diagnosis. According to the National Institutes of Mental Health, an estimated 17 million Americans, about 7 percent of the population, had a depressive episode in 2017. Clinical depression is even more common in younger adults, affecting more than one in ten 18–25 year-olds. Estimates of the costs of depression in the United States vary, but some put it as high as $210 billion annually, when we factor in costs of treatment and loss of productivity from missed days or underperformance at work. Globally, depression is estimated to affect over 300 million people and is listed by the World Health Organization as the single largest contributor to disability.[30]

The incidence of depression has been rising during the digital age. A detailed analysis of data from 2005 to 2015—a time of explosive growth in smartphone and social media use—finds that the incidence of depression among US adults rose from 6.6 percent to 7.3 percent during the period, which translates to over 100,000 more cases. The rate of increase was higher still among adolescents (ages 12–17 years), with prevalence rising from 8.7 percent to 12.7 percent.[31]

Globally, we see similar, and sometimes worse, trends. According to the World Health Organization, during the same 2005–2015 period, depression rates around the world rose a staggering 18 percent. "These new figures are a wake-up call for all countries to rethink their approaches to mental health and to treat it with the urgency that it deserves," said Margaret Chan, former director-general of the World Health Organization, in response to the increase.[32]

Another indicator that depression is on the rise? According to the National Center for Health Statistics, the use of antidepressants skyrocketed 400 percent between the early 1990s and the mid-2000s. While many of the uses of modern antidepressants go beyond depression and include anxiety disorders, neuropathic pain, eating disorders, and premenstrual syndrome, the rise is certainly noteworthy. Clearly something is changing the minds, brains, and moods of an awful lot of people.[33]

It is not entirely the fault of social media companies that so many bad experiences occur on their platforms, potentially resulting in serious mental health consequences—although the companies should be held accountable for knowingly and deliberately designing algorithms that promote conflict, fear, and outrage among users. Still, it behooves users to limit their time on social media and to try to spend that time more wisely. Social media ultimately is a tool for amplification. It will take what we give it and give us back as much as possible of the same thing—over and over, via intrusive push notifications, recommended ads and posts, autoplay, and infinite scrolling. In this respect, social media can be a depressed person's worst nightmare, providing endless grist for ruminations.[34]

Those Who Compare Will Despair

If you want to feel bad, just spend a few minutes thinking about how much better off your friends are. That's what Jennifer Garam, our blogger and yoga fan, discovered after spending time on Facebook. Not that we needed social media to realize this. Edith Wharton knew it. The author of *The Age of Innocence* prominently featured portraits of affluence in her celebrated works, inviting social comparisons in the 1920s. Some have suggested that the phrase "keeping up with the Joneses" refers to Wharton's birth family—an upper-class New York clan known for ostentatious displays of wealth that spurred competition among their high-society neighbors. But social comparison isn't just for the rich. Since at least the 1950s, social psychologists have theorized that people have an innate desire to measure the quality of their lives against the perceived quality of others', giving rise to a field of academic study known as social comparison theory.[35]

Social comparison is not just a fancy synonym for envy; one can experience a range of emotions, depending on one's relation to the object of comparison. For example, when we compare ourselves to someone we think is superior in position—an upward social comparison—we may experience envy. But we may also have a positive emotional response, feeling inspired or optimistic. And when comparing ourselves to someone we think is inferior in position—a downward social comparison—we may feel pride or sympathy.[36]

Research on social comparisons offline has shown that when we see ourselves as inferior to others, we are more likely to judge ourselves harshly and experience negative emotional effects such as depressive symptoms and reduced self-esteem. There is some evidence that the inverse is also true—that when we see ourselves as superior to others, we experience higher self-esteem and lower anxiety. There is also research suggesting that it is not the types of social comparisons we make but rather the frequency of social comparisons that leads to negative consequences.[37]

Social media takes this common tendency to compare ourselves to others to a whole new level, with consequences for mental health. A 2018 study of South Korean Facebook users found that psychological well-being was directly tied to the type of social comparisons they made on the platform. Users who experienced social-comparison posts as inspirational or evoking sympathy had improved well-being. In contrast, users who experienced social-comparison posts as evoking envy or depressed feelings had reduced well-being.[38]

Mai-Ly Steers was a postdoc at the University of Houston in 2014 when her younger sister inspired her to investigate the consequences of online social comparisons. One weekend, her sister was home and feeling blue because she was not invited to a school dance that evening. Feeling left out, she decided to check Facebook to see what she was missing. This turned out to be a big mistake. "She saw friends posting pictures from the dance and felt really bad," Steers told the *Washington Post*. "She was gathering info about the dance that she wouldn't even have known about [had she not checked Facebook]. I began to think this is probably a common occurrence."[39]

Steers carried out two studies to learn more about the amount and type of social comparisons we make online. Interestingly, she found that negative mental health outcomes were associated with the number of online social comparisons users made, regardless of the type of comparison. It did not matter if the comparison was upward, downward, or nondirectional. The more time her study participants spent on Facebook, the more social comparisons they made, and higher quantity of social comparisons was significantly related to depressive symptoms.[40]

Garam offers one possible solution to mental health problems that many people of all ages experience when using social media: get offline. "If I want to have any sense of self-esteem, I have to pull my gaze off the social media streams and place it firmly back on my life," she writes. "I can't linger on social media; I have to check it quickly, sideways, and with my eyes squinted, and when I see something that triggers thoughts of not measuring up and feelings of being a failure, I have to get off, STAT."[41]

This is good advice. Social media researchers are almost always careful to point out that the ill effects of online social networking are not necessarily intrinsic to the platforms. Yet the platforms are designed to amplify what users are thinking, feeling, and experiencing, potentially intensifying our worst behavioral tendencies. As a result, more time on social media tends to mean more low-quality experiences and interactions, more exposure to negative stories and images to ruminate on, and more of the sorts of invidious social comparisons that bring us down. Social media may not often lead to anxiety and depression among those who aren't predisposed, but those who are predisposed are at considerable risk. The offline world has its share of stimuli, but the online world is an unrelenting, concentrated stream that threatens to pull under those of us already on shaky ground.

Increasing Narcissism

In Greek mythology, Narcissus was a man of extraordinary beauty, pride, and arrogance. In one version of the story, Narcissus displeased the goddess of revenge, Nemesis, by spurning the love of a nymph named Echo. The young woman was distraught over the rejection and never recovered. In response, the angry Nemesis punished Narcissus by luring him to a pool of water when he became thirsty after a day's hunt. As he stared into the pool, he admired the beauty of the reflection and, never realizing he was looking at himself, fell deeply in love with it. But after a time, his love turned to suffering, as he realized that his affection would never be returned—just as he had not returned Echo's. In some versions of the story, Narcissus takes his own life in despair. In most, he turns into a small white flower, which to this day is known as a narcissus. His other legacy, inspired by his lack of empathy and abundant but shallow self-love, is the term "narcissist."

You have met people with narcissistic tendencies or downright pathological narcissism, clinically known as narcissistic personality disorder. Like Narcissus, they tend to admire themselves. They also tend to be grandiose, extroverted, overly confident people with a strong sense of entitlement, an inclination toward antisocial antics, and an insatiable need for flattery. Many narcissists lack empathy for the suffering of others. Clinically, narcissism also has a more introverted subtype. This subtype is exemplified by low self-esteem, self-hatred, and high anxiety that can result in social isolation. Common to both types of narcissism is a constant self-absorption that interferes with relationships. While many narcissists appear to have normal, "functional" relationships, these relationships are often quite shallow and serve mainly to support the narcissist's ego.

While there is some debate among psychiatrists about how best to characterize narcissistic personality disorder, one definition that unifies both types of narcissism includes a "fragile sense of self that is predicated on maintaining a view of oneself as exceptional." This definition is appealing in part because it directs attention both to the brittle nature of narcissistic personalities and the resulting need for external feedback: narcissists lack the internal mechanisms to achieve self-esteem, so they look for attention, support, and validation from the outside world. What better platform could there be to receive that external feedback than ubiquitous social media apps?[42]

The first study to suggest that social media networks may be fertile ground for narcissists was published in 2008. The assumption was that social networks offer narcissists easy access to a large audience and nearly instantaneous feedback. The study assessed the Facebook pages of a group of users alongside their scores on a widely used clinical scale called the Narcissistic Personality Inventory. Independent raters coded the users' Facebook pages for a variety of factors that the authors surmised would be related to narcissism. The results showed that users with high narcissism scores tended to have more people in their networks and more self-promotional material in their personal descriptions. The raters also gave these users' profile photos higher marks in terms of "attractiveness, self-promotion, and sexiness." The authors concluded that narcissism is readily identifiable online and is portrayed there in much the same ways as it is in the offline world.[43]

So the intuition that social media is a good outlet for narcissists is probably sound. But the relationship between narcissism and social media may run deeper. A key question is whether narcissists have a special attraction to social media and whether social media may contribute to increasing narcissistic tendencies over time. There is some evidence to support the idea that Americans are becoming more narcissistic.

Jean Twenge, the psychologist who studied generational shifts in teens and related them to changes in online behaviors, reports that the Narcissistic Personality Inventory scores of college-age Americans increased significantly between 1979 and 2006. Almost two-thirds of those in the later part of the sample scored above the average of the earlier sample cohorts.[44] While the range of the study is impressive, subsequent research on narcissism has been mixed, and 2006 was a long time ago in social media years.[45] What does more recent research tell us?

As is often the case, a literature review can help us understand whether social change is really happening. A 2018 meta-analysis sheds light on the subject. The research brought together fifty-seven studies of narcissism involving over 25,000 social media users from around the world. The results suggest a statistically significant relationship between "grandiose narcissism" (the more extroverted type) and social media use. In particular, narcissists tend to spend more time on social media, have larger numbers of contacts, upload more photos, create more posts, and comment more on others' posts. Findings across the fifty-seven studies differ somewhat with nationality but not with age, gender, or year of publication. Nor does the specific platform appear to matter. Narcissists seem to be equally at home on Facebook, Twitter, and other services.[46]

These findings make abundantly clear that narcissists make greater use of social media, but again, we are left with the question of causality. Do narcissists seek out social networking sites, or do social networking sites cause or reinforce narcissistic tendencies?

The authors of the meta-analysis suggest that both may be true. Just as those prone to depression are more likely to have negative ruminations and share them online, individuals predisposed to narcissism are more likely to engage with social media platforms, and their engagement with those platforms reinforces their tendencies. If, as several researchers suggest, narcissism is a spectrum disorder, with many people showing some tendencies and fewer showing full-scale clinical features, then social media might be a perfect means for encouraging symptoms of narcissism to blossom. The logic is straightforward: those who post more frequently and have more contacts will reach more people and thereby generate more responses. This is, in and of itself, a feedback loop that encourages narcissistic extroversion. And social media algorithms will tend to intensify that feedback, as posts that generate high response rates receive priority, showing up in more feeds for longer periods of time. So there are clear mechanisms whereby online narcissism begets more narcissism.

This is alarming for at least two reasons. First, because the structure of social media rewards narcissism, we will, over time, probably be exposed to more narcissistic people and their posts online, inviting more social comparisons. Second, rising rates of narcissism contribute to lower empathy, less healthy relationships, and higher rates of depression across society, all of which we are in fact seeing.[47]

More research on social media and narcissism is needed, but when we look at available findings in the aggregate, the trends are concerning. As the distinguished psychiatrist Ravi Chandra puts it in a column for *Psychology Today,* "All the world's religions are essentially aimed at transcending self-centeredness. Social media, on the other hand, can be seen as a Temple of the Self." Twenge worries that "the dark side of individualism" is taking over—"freedom without responsibility, relationships without personal sacrifice, and positive self-views without grounding in reality."[48]

If indeed we are seeing a higher incidence of narcissism encouraged by online feedback loops, the results will get worse over time. The self-focus inherent in narcissism makes for reduced empathy, which fuels intolerance, political polarization, and dehumanization. Vulnerable minority and immigrant groups are likely to feel the brunt of empathy deficits. And stemming the problem is difficult because there is no agreed-upon treatment for narcissistic personality disorder. We also know that narcissistic personality types respond less successfully to treatment for other mental health problems such as depression and anxiety disorders. A concerning trend indeed.[49]

The neurological correlates of narcissism provide yet more reason for worry. Researchers in Germany studied adults with clinically defined narcissistic personality disorder and compared their brains with healthy adult brains matched for age, gender, and intelligence. The researchers found that the narcissistic adults had smaller volume (less gray matter) in their dorsolateral and medial prefrontal cortex. They also had less volume in an area of the brain known as the anterior insula.[50]

The findings of a smaller prefrontal cortex that is less capable of much-needed orchestration are concerning for a number of reasons, as we have seen through many examples. As for the anterior insula, this brain region has been implicated in empathic abilities; studies show that a smaller and less active anterior insula correlates with reduced empathy. Consistent with this finding, looking across both narcissists and healthy controls, the German researchers found that lower empathy scores were associated with a smaller anterior insula. The results suggest that the brains of narcissists are fundamentally different than those of non-narcissists in key areas related to empathy, emotion regulation, and attentional control.[51]

The jury is still out on whether social media, in general, has been good for its billions of users individually and for society at large. But evidence is mounting, and we now know enough that we should recognize the need for caution. Another way to put this is that we should all be able to agree that social media, like any tool, can be used for good or ill. Social media can enhance offline social bonds and can be used to create new bonds, including meaningful connections that play out exclusively or primarily online. In these respects, social media, and the devices we use to access it, can be empowering. However, if the mere presence of those devices interferes with our social bonds, or if we use social media to make unfavorable social comparisons, perseverate over negative thoughts and feelings, and feed our self-centered tendencies then we will have lost control of our tools. Instead of using these tools to improve our lives and our communities, we will develop lower self-esteem, less empathy, and more depression and narcissism.[52] This leads to our final consequence for adults.

An Epidemic of Loneliness

John Cacioppo was a prolific psychologist and neuroscientist at the University of Chicago. He was the father of social neuroscience, the field I entered early in my career as a young physician-scientist studying the neurobiology and physiology of empathy in the clinical relationship. I cited his work often and was thrilled to have lunch with him one time in Cambridge, Massachusetts, before he delivered a stunning lecture at the MIT Media Lab. He was kind, considerate, encouraging of my research, and extremely warm in his demeanor. Late in his career, he turned his extraordinary intellect to the impact of human social relationships on our biology and our health. Before his premature death in 2018, he wrote a book on the topic of loneliness.

Coauthored with science writer William Patrick, *Loneliness: Human Nature and the Need for Social Connection* came out in 2008, at the dawn of the smartphone revolution. The authors take an evolutionary perspective on socialization and loneliness, arguing that humans are not the rugged individuals of American lore but rather the product of our relationships. Historically, loneliness has often been viewed as a facet of depression, but this misses the essential point. If lonely people are depressed, this is because loneliness itself is fundamentally at odds with human well-being.[53]

As we saw in part 1, strong social bonds in infancy are essential to development, and strong bonds in early childhood are the foundation of healthy attachment styles as we age. Relationships are also essential to our species, as nurturing relationships gave rise to the family unit, and groups of families gave rise to communities. Communities expanded through shared language, culture, and history, forming the basis of modern societies and politics. For the entirety of our time on this planet, humans have depended on strong social bonds. They are the basis of our complete dominance of the earth, for better and for worse. The effects of our social relationships are profound. They play a role in our perceptions of the world, our behaviors, and ultimately our brain physiology— even affecting our DNA. We are wired to connect.

But in recent decades, we have departed radically from our evolved social habits. For example, the prevalence of one-person households in the United States has increased dramatically over time. In the early 1900s, about 5 percent of US households comprised one person. By 1940 the rate had risen only modestly, to 8 percent. By 2013 the proportion had increased to 28 percent, suggesting a significant rise in social isolation over the second half of the twentieth century.[54]

People who live alone are not necessarily lonely, but there is evidence that we are increasingly feeling the sting of isolation. In 1980 scholars estimated that about 20 percent of Americans felt lonely at any given time. By 2012 that number had doubled to 40 percent. A 2018 study of nearly 20,000 US adults put the number at 46 percent, with Gen Z young adults reporting the highest levels of loneliness. Loneliness has gotten so bad that Vivek Murthy, surgeon general in the Obama and Biden administrations, ranks it as a growing threat to national health and well-being. "As a society, we have built stronger Wi-Fi connections over time, but our personal connections have deteriorated," Murthy mused in 2017.[55]

Cacioppo describes how loneliness is like hunger, thirst, and pain in that it has evolved to tell us we need to do something in order to take care of ourselves. Loneliness triggers aversive signals in our brains as a reaction to the threat posed by social isolation. These aversive signals motivate us to change our behaviors in ways that return balance to our bodies and our brains—what physiologists call homeostasis. We evolved to be socially connected, so when we don't have strong social bonds in our lives, we feel displaced or depressed—a sign that we are lacking social balance.

Loneliness, in other words, is an indicator that our social self is suffering. "Home sickness, unrequited love, bereavement, being shunned or put down," all of these feelings, Cacioppo states, are part of the aversive signaling that something is wrong. When we feel lonely, we think we are in trouble, and we snap into self-preservation mode. Lonely people have higher levels of stress hormones, a sign that their brains and bodies perceive threats to survival. Brain-imaging studies of lonely individuals show shifts in activity away from empathy centers and toward the visual cortex, another evolved maneuver that indicates a perception of threat. Lonely people are on the lookout for hazards in their environment. They probably won't find these hazards, but that is not the point. The point is that when we are lonely we are wired to perceive the environment as threatening, and our brains and bodies respond accordingly.

These deep evolutionary responses play out below our conscious awareness and push us to change our behavior in order to protect ourselves. In the short term, the problems often go away, either because we solve them proactively or take control of our thoughts and realize that maybe the situation isn't so bad after all. But what happens when loneliness becomes chronic or can't easily be resolved? Loneliness can be self-reinforcing in much the same way as depression: when we are on high alert and looking for dangers, we are more likely to see danger, leading to patterns of self-defeating thoughts. The stress response itself, if prolonged, leads to a host of other mental and physical problems over time.[56]

Chronic loneliness and its associated stress reduce sleep quality, degrade physical health, make us more prone to depression and addiction, and can even lead to premature death. Yes, strong social bonds are so important to our survival that their absence is a potent predictor of shorter life expectancy. In fact, chronic social isolation is more dangerous than obesity and other behavioral risk factors such as lack of exercise. Some experts put the health risk of chronic loneliness as the equivalent of smoking fifteen cigarettes per day for life—a 20 percent increase in risk of premature death. Meanwhile, other data show that people with strong social bonds are 50 percent less likely to die over a given period of time than those with fewer strong social connections.[57]

Social homeostasis, an evolutionary adaptation acquired over a million years, is misfiring in the digital age. Media technology is not the only factor contributing to increased social isolation, but it is easy to see how the tools at our disposal are at odds with our need to connect. Consider one subtle but powerful change in our own living rooms brought on by new media and the digital world: the switch from "appointment" television viewing to on-demand video. For decades television content appeared according to a strict programming schedule. Anyone in your household who wanted to see a particular show had to come together to watch it. Appointment television also enabled the so-called water cooler effect that many of us grew up with—gathering with friends and colleagues at school or work to discuss the show we all watched last night.

Such shared experiences are much less common today. A plethora of new distribution channels, more access to broadband and on-demand programming, and the extraordinary profusion of content created by professionals and amateurs alike mean that viewing has become fragmented. In 2002 US television broadcasters aired 182 scripted programs; in 2016, the number had more than doubled to 455. And instead of talking about shows with friends, we started commenting on them on social media and on blogs. In large part that is because we are watching on our own schedules, streaming online and "binge" viewing on laptops, phones, and tablets.[58]

If this seems like an odd complaint, keep in mind how much time Americans spend watching Amazon, Netflix, Hulu, and other streaming services, alongside on-demand programs from traditional broadcasters via smart TVs. Television was perhaps never the ideal social-bonding tool, but it was for decades at least able to create some sense of shared culture and experience. The many hours we spent watching traditional TV created common experiences that inspired conversation topics that brought people together. That hardly exists anymore, undermining one kind of opportunity for social bonding and meaningful face-to-face interactions, contributing to our loneliness.

There is even evidence to suggest that as we have become more socially isolated, we are also becoming more comfortable interacting with media and other technology than we are interacting with each other. This was the main finding from a chilling 2014 publication by researchers at Dartmouth College and Harvard University. The researchers gathered two groups of participants; one group was socially disconnected (lonely), while the other enjoyed healthy social networks. All participants then viewed a series of images of human faces, some of which were digitally morphed so that they appeared inanimate and doll-like. The images ran the spectrum from truly human to not-at-all human-looking.

The researchers found that the more lonely and disconnected individuals were, the more likely they were to see human faces in the images digitized to appear inanimate. The authors conclude, "In their efforts to find and connect with other social agents, individuals who feel socially disconnected actually decrease their thresholds for what it means to be alive, consistently observing animacy when fewer definitively human cues are present."[59]

As we increasingly substitute computer-mediated relationships for in-person ones, we degrade our social skills and connections, with real consequences. Repeated, impulsive, habitual, and sometimes addictive use of mobile media, communications, and information technology is self-fulfilling. It rewards us for rewiring our brains in the direction of distraction, empathy deficits, depression, narcissism and ultimately loneliness. Those predisposed to mental illness are in the greatest danger, but all of us run the risk of feeling lonely and lowering the threshold of what we perceive as human. Is there anything we can do to stop these trends? It is time to become more proactive than reactive. Part 3 begins a journey toward a new tech-life balance.

PART III

BEYOND WIRED

Better Brains

8

FIRST, RECOGNIZE THE PROBLEM

Your friend has been acting a bit strange lately. At first it was hard to put your finger on what exactly had changed. It was a subtle shift that took place over the course of months, maybe even years. But at some point you began to notice moments of unexpected irritability. She was getting frustrated easily, and seemed less empathic, less dependable. There were periods of social withdrawal, and when you were together, her attention would drift toward her smartphone. She was always on it, even when she was driving or when her kids were trying to talk to her. What really tipped you off was how nervous she got when her phone battery was low, which happened constantly because she would never put the thing down. And there was that time when you tried to go for a hike together, but it turned out she had left her phone in the car. Anxiety leaked out of every pore, and she became defensive when you pointed out that there was no need to go get it.

Recognizing problematic behaviors in ourselves and those around us is never easy. And recognizing problematic technology behaviors is especially challenging because all of us are accustomed to seeing smartphones, tablets, and laptops everywhere, in every context—while comparatively few people realize that it's even possible to have a troubled relationship with modern media, communications, and information tools. One way to sensitize our communities is to raise awareness of the risks. That is the goal of this book.

Individually, we should keep our eyes open for warning signs, which often take the form of a noticeable change in behavior. Are you or is someone you care about suddenly spending more time immersed in a smartphone, app, or game? Do your young children have temper tantrums if you withhold or limit screen time? Is your tween or teen child using social media more frequently than they used to, or texting from unusual or inappropriate settings—maybe from the dinner table, the bathroom, or during a conversation with you? Are unexpected packages arriving from online retailers, filling your house with unneeded clothing, gadgets, and appliances? All of these could be signs that you or someone around you is losing control of their online habits.

Behavior changes are often subtle and may appear only as indirect evidence of the underlying problem. For instance, you may notice that your teenager never seems to get enough sleep, even though he is always in his room at bedtime. Or maybe he is unexpectedly spending a lot of time in his room in the middle of the day. Perhaps you've noticed that your coworker lingers in the bathroom on a regular basis or comes into work looking burned out. All of these could be signs of a problem, including potentially a bad technology habit or addiction. Intervening in cases like these is challenging, but detection is the first step toward a solution, and the earlier the detection the higher the probability of successful resolution.

A sure sign of problematic technology use is the inability to cut back. A person may recognize that they are hurting themself or others and still fail to change their behavior, even after multiple attempts. It is so easy to delete an app and declare, "Never again!" before reinstalling it a few days later. Even a brief detox can be revealing, though. If a concerted effort to limit use results in withdrawal symptoms such as irritability or anxiety, then there is little doubting the seriousness of the situation. Withdrawal symptoms do not definitively reveal addiction, but the association is strong.

We should be especially vigilant with friends and loved ones who have mental health problems or a family history that predisposes them to mental illness. Problematic technology use may worsen symptoms of an existing mental health problem. Meanwhile, those with a history of hostility or aggressive behaviors, depression, anxiety, substance abuse, or ADHD, and those who have struggled with trauma are at greater risk of developing problematic technology behaviors. All mental illness affects our prefrontal cortex to some extent, so it is not surprising that mental illness also would increase the risk of becoming a compulsive or problematic technology user.[1]

Of course, like smartphone and internet addictions, other mental illnesses are often suffered in silence, the symptoms readily masked or hidden. For this reason, it is essential to practice compassion and empathy even when there is little reason to suspect a problem. If we want our friends, coworkers, and loved ones to be open with us about the difficulties they face, they need to feel secure in sharing. We create welcoming environments through routine behaviors of care. Even then, however, struggling people don't necessarily come forward. If you suspect someone is suffering or showing early signs of a problem, don't ignore them. Ask questions, educate yourself using credible resources, and seek medical advice.

Warning Signs of Problematic Smartphone Technology Use

1. Increased time spent with technology
2. Increased time spent alone with technology
3. Preoccupation with a particular app or device
4. Loss of interest in other activities
5. Increased irritability, anxiety, or depressed mood
6. Attempts to deceive or hide use from others
7. Unsuccessful attempts to cut back or change behaviors
8. Signs of withdrawal when technology is not available
9. Interferes with social interactions and meaningful relationships
10. Interferes with school or work performance

The clearest indication of an unhealthy habit or addiction is that important aspects of a person's life are starting to unwind. Often these indicators arrive too late, after the problem is well established and therefore harder to tackle. When children are in the midst of technology addiction, they typically will retreat into isolation and become combative in interactions with friends and family members. Their academic performance will decline, and they may have other problems at school, such as tardiness and falling asleep in class. For their part, adults with addiction experience frayed relationships, as they ignore friends and loved ones. Adults tend to become less productive at work, whether because of distraction or fatigue. They may show up late or fall asleep on the job, owing to late nights spent watching, gaming, and posting online. In extreme cases, adolescents and adults may turn to alcohol, marijuana, or other drugs to self-medicate and escape the anxiety and depression that result from withdrawal or excessive technology use.

Is It Healthy?

Mental health experts spend a lot of time discussing and studying the signs, symptoms, and neurobiology of pathological brain states. But we sometimes get so focused on the definition of what is unhealthy that we forget to ask a no less important question: What does a healthy psychology look like?

Sigmund Freud, the founder of psychoanalysis, was supposedly once asked a version of this question and replied that a healthy person should be able "to love and to work." Another take on the phrase, which I learned during my psychiatric training, is more effusive: "Love and work are the cornerstones of our humanness." It is not clear that he ever said these words, but they provide a simple benchmark for mental health. According to one interpretation of this sentiment, a mentally healthy person maintains at least a few meaningful, reciprocal social relationships—Dunbar's small circle of intimates—and is able to do something productive with their time. This might be paid or unpaid work, and in the case of children, this typically means schoolwork, sports, or some other extracurricular activity. If either of these aspects of life goes astray—if one cannot maintain a few strong social bonds or cannot engage in meaningful work of some type—that may be a sign of problematic habits or addiction.

Love and work require a healthy prefrontal cortex enabling emotion regulation, planning, and focused attention. We can and should aspire to more. But as long as we're all using our smartphones every day at work, school, and in our personal lives, we need some minimum standard for a healthy tech-life balance. So ask yourself this: Is technology helping you on your journey, or is it standing in your way? We should all be able to point to at least a few fulfilling relationships of which we are a part. We should all be able to say that our work makes use of our skills, adds knowledge and meaning to our lives, or contributes to the betterment of our communities, however defined. Is your relationship with technology a healthy one? It is if it supports these aims.

Create Goals

When we feel the pull of an app ping or message ring, the reward of checking our phone is immediate, even if it comes at some cost—a disrupted conversation with a friend or distraction from the task at hand. Holding off the urge to check and instead being present with your friend or staying on task doesn't always yield such immediate rewards. Instead, the payoff of strong relationships and successful completion of work and school assignments often comes down the road. Most of us prefer gratification now to success years from now. One way to check ourselves, to set aside temptations without hard feelings, is to have a goal in mind. What are you trying to achieve by putting off your immediate reward, and how will you know you've achieved it? Goals enhance motivation and inspire action. This is particularly important as we think about changing habits or creating new habits whose benefits may not be apparent for some time.

Behavioral psychologists have set out several steps to proper goal-setting that can help us with the formation of new habits. First, the goal needs to be clear and framed properly. When you are trying to do something very hard, like overcoming a strong habit or an addiction, you are asking your brain to set aside what it obviously wants in exchange for what seems like a competing outcome. In the moment, the pull of the reward centers—the craving—is many times stronger than the small satisfaction that comes with knowing you are working toward regaining control. To overcome this challenge, we need to set goals in ways that improve our chances of success. To this end, researchers have found that framing goals in positive, outcome-centered terms tends to be most effective. Instead of telling yourself what you want to avoid—"I want to check my smartphone less when I am around friends and family"—tell yourself what you want to achieve: "I want to have better relationships and be more productive at work."[2]

Second, be mentally aware of your goals and whether your present actions comport with them. While much of habit formation happens without our conscious awareness, we can be mindful of our actions and intentions. Mental awareness is key to forming new habits because we need to do the desired action repeatedly in order to create lasting behavioral change. Repetition exposes us to the contextual triggers and environmental situations that reinforce the new habit. In time, repetition retrains the reward centers in the brain. Eventually the striatum will kick in, and we will not need to consciously control the new action we wish to take. At that point a healthier habit will replace a destructive one, and we will have succeeded in achieving our goal. But at the beginning of a process of behavioral change, the new habit isn't a habit; it isn't automatic. So, in the early going, we have to rely on our prefrontal cortex to monitor our situation constantly and intervene on the side of our better judgment. This is the essence of mental awareness.

Third, begin pursuing your goal on a date that is meaningful to you. It turns out that temporal landmarks help us achieve goals, which helps explain the popularity of New Year's resolutions. Research shows that starting our pursuit of a new goal on our birthday, the first day of a vacation, immediately following school graduation, or upon some other milestone date motivates us to take up the challenge. This is known as the fresh-start effect.[3]

Fourth, we need to be mindful of scenarios that can set us back and be prepared for them. There are many situations that diminish our motivation to pursue future rewards and instead push us toward the default behavior we are trying to avoid. Stress, negative emotions, and ubiquitous temptations test us. When we are faced repeatedly with these daunting obstacles, our motivation and self-control can wither. Like a muscle fatigued by overuse, the prefrontal cortex can shrink from the challenge of overcoming unhealthy habits, diminishing our resolve. This is why when we are tired, stressed, or otherwise depleted, we fall back on old ways. Obtaining the immediate rewards of an unhealthy habit requires much less mental effort than creating a new habit in hopes of experiencing rewards later.[4]

While we shouldn't give ourselves too many passes, we do need to recognize that kicking an unhealthy habit or addiction is going to be hard, and our efforts will lapse at times. This is normal and to be expected. Breaking an unhealthy habit means overcoming the large neurological investment we have already made in it. Mobile media, communications, and information technology habits are based on behaviors repeated hundreds of times every day and rehearsed for years. Change will be difficult, despite our best intentions and even if we have clear, well-framed goals.

It is no wonder that studies show that less than half of New Year's resolutions are still being pursued a year later. Indeed, the chance that an unhealthy habit will get the better of us is quite high. A review of more than sixty studies on habits and goal intentions found that habits often persist in the presence of competing goals. That is, existing habits most often win out against our best intentions.[5] There is even research that suggests that activation of the habitual response—say, checking your smartphone in front of friends or while driving—reduces our mental access to alternatives, such as the intention to put and keep the phone away. In other words, strong habits force us to process information in a way that will decrease the likelihood of considering alternative behaviors based on new goals or intentions.[6]

The good news is that what is hard at first eventually becomes easy: with time and intention, a beneficial habit can be as accessible to our brains as an unhealthy habit is. While building a new habit, we need to be conscious of not lapsing into old ways even as our stressed and worn-out selves wish to simply relax and do what comes readily. But once we have made the shift to healthier habits, those are what will come readily. Overcoming unhealthy habits takes intention, willpower, clear and well-framed goals, and often the support of friends, family, and caring professionals. But over time, exercising healthy habits requires none of these. For a person with solid tech-life balance, ignoring an app ping or message ring is automatic.[7]

Addiction to cigarette smoking provides an illustrative parallel. The ills of cigarette smoking are no secret these days. Smoking leads to disabling lung and heart disease, cancer, and early death, and secondhand smoking imposes health hazards on others. Most smokers know this, which is probably why the vast majority want to quit. But there is a reason why 90 percent of smokers who try to quit on their own initially fail. Nicotine and the behaviors associated with smoking are powerfully addictive, and the cravings that drive smokers to light up arise in response to many triggers in the environment, including stress and negative emotions.[8]

However, nearly 40 percent of smokers can quit for meaningful periods of time when they employ strategies that go beyond the use of willpower alone. These strategies include setting a clear date to quit, getting social support, and discarding paraphernalia associated with smoking. These strategies are most successful when combined with so-called temptation bundling. Temptation bundling is a method of leveraging the power of an unhealthy habit in order to foster a healthy one. Basically, we try to associate an existing reward pathway with a new, nonharmful stimulus. Some smokers might collect the money they would otherwise spend in a jar so they can watch it accumulate. The sight of that mounting pile of cash can be a strong motivator; money, after all, stimulates the same reward centers as drugs. Or a smoker might train himself to squeeze stress balls or suck on lollipops whenever cigarette cravings arise, in order to perform habitual hand and mouth rituals through other means. Some smokers who wish to quit motivate themselves by visualizing a distant future in which their children or grandchildren have grown up—a future they might not be around for if they remain addicted to nicotine.[9]

Most smokers will find that the people around them strongly support their goal of quitting. This is one way in which nicotine addiction can differ from tech addiction, and one reason that tech addiction can be even harder to overcome. As noted before, tech addiction and unhealthy smartphone habits are often unseen or are easy to hide, so our friends and loved ones may not realize we are struggling. And where tech addictions are isolating, they will tend to reduce the size and reliability of our support structures. This is tragic because social support is critical to realizing the goal of tech-life balance. The fact is that most of us need mobile media, communications, and information technology in our lives; we cannot go cold turkey. So there will be lots of opportunities to slip on the path to recovery. Following that path will be easier if we can enlist friends or family members. These accountability partners help us see the value of achieving our goals and commit to assisting us along the way. When we struggle to see how tomorrow's rewards outstrip the immediate gratification available to us today, an accountability partner can lend their prefrontal cortex to us, refocus us on how much we stand to gain, and help us establish control in the moment.

Finally, in order to achieve effective habit change, we need to be modest with our goals and accept our limitations. The experiences of extreme athletes are revealing here. Research shows that many of them start off like you or me, but they diligently monotask with repeated, effortful practice and focused attention. And they push a little each time, often in 4 percent increments and in line with their challenge-to-skill ratio. In other words, an extreme athlete isn't some average Joe or Jane who sets a goal to wake up tomorrow and run an ultramarathon, scale tall mountains, or hold their breath for an insane period of time. Extreme athletes incrementally expand their capacity a little bit at a time, day after day, as their skill level allows. Our ability to focus and achieve maximal productivity is dependent on a very specific relationship between the difficulty of the task and our ability to perform the task. If the challenge is too great, fear and anxiety overwhelm us, and we reach for an alternative. If the challenge is too easy, we get bored or distracted and, again, reach for an alternative.[10]

Like a marathoner's leg muscles, the human brain can fatigue, but its capacity can also be expanded with considered effort and repetition. We make more neurons (neurogenesis) and new connections (neuroplasticity) every day. We are constantly rewiring our brain, for better and for worse. Change is possible, but it won't happen overnight. Setbacks are inevitable. Like the elite athlete who painstakingly practices a little bit each day, we need to learn to focus and stay in the moment while we develop new habits that could produce some future reward.

In the next chapter, I offer ten scientifically sound recommendations for creating a healthy tech-life balance and a new digital literacy. In each case, I point to research that demonstrates the long-term value of following the recommendation and achieving the goal, the better to motivate readers. These are ten strategies that go beyond willpower. They are not the trite tricks or "brain hacks" you may read about in other books or online. All of the recommendations are designed to strengthen the prefrontal cortex in service of supporting the conductor of our own personal symphony.

9

TEN RULES FOR A
HEALTHY TECH-LIFE BALANCE

This book is heavily focused on the challenge of realizing personal well-being in the digital age. However, as I turn to the recommendations, I want to suggest a further motivation for changing your relationship with mobile media, communications, and information technology. A healthy prefrontal cortex is indeed good for you—it means greater productivity and sturdier social connections. But these are also good for society. We will need strong communities, full of capable, mature people at their best in order to tackle the many social, political, and environmental challenges we face—climate change, pandemics, political divisiveness, racism, and income inequality, to name a few of the most pressing. Improving brain health, reducing the toll of mental illness, and fostering better relationships will improve our collective welfare. But to get there, we need clear rules—some speed limits and signposts on that growing and increasingly complex information superhighway. We can learn not to scratch every itch, not to succumb to every app ping and message ring.

#1 Stop Multitasking

Imagine you could quickly increase your productivity, improve your concentration, and connect with friends and family in a more meaningful way that ultimately leads to a happier and more fulfilling life. Does it sound too good to be true? In fact, there is a clear remedy for some of us. Decades of research show that multitasking reduces our processing speed, accuracy, and productivity. Heavy media and technology multitaskers are more distractible and have more trouble switching between tasks efficiently compared to their light multitasking peers. Heavy multitaskers are more likely to succumb to filter failure and automatic attention.

In order to stop this destructive habit, we need to be mindful of the many inducements to multitasking. Calendar alerts, instant messages, and a range of notifications are all common in the workplace and increasingly common for students. They all contribute to multitasking. A classic Microsoft study found that employees working independently at their computers spent an average of nearly ten minutes with a secondary task when triggered by an alert that interrupted their primary work goal. In many cases, these Microsoft employees were distracted by interrupting tasks for more than two hours per day.[1]

Perhaps because this was Microsoft research, the study authors suggested a number of software solutions. For instance, they found that when multitasking, if users could still see the window or application that contained the original suspended task, they resumed that task more quickly—so one suggested fix is to incorporate software on work machines that reminds users to return to their original work interface after a period of nonuse. For most of us, a better strategy is behavioral: when engaging in important or cognitively demanding tasks at work or for school, close any documents, email, and calendar programs and any other applications that may send notifications or automatic alerts.

There is good news for those who seek to change this habit: the rewards of monotasking are considerable, measurable, and almost immediate. When switching from multitasking to monotasking, we will quickly find that we are more focused, more productive, and more attentive to others. If you are on your smartphone, tablet, or computer and you are doing more than one task, take a moment to pause and ask yourself: Is it worth it? Are you compromising some other more important goal, relationship, or indeed your safety or the safety of others? Mental awareness in this case is situational awareness. We need to remind ourselves of the consequences of multitasking and of the reward of increased focus, productivity, and stronger social bonds—rewards that we will be able to realize as soon as we are doing just one thing at a time.

The new habit of monotasking means we put our smartphones away while studying; close email, social media, and news feeds at work; turn off notifications and keep our phones out of sight when spending time with friends and family. This does not mean using mobile media, communications, and information technology less, although in many cases that may result. What it does mean is using them more intelligently.

Research suggests that many of us cannot rely on willpower alone to prevent multitasking. We have built so many habits around the myriad temptations that come from the supercomputer in our pockets that we cannot go it alone. So we can respond by removing some distractions. That's what turning off notifications does. We can also begin to form our own digital etiquette and ask our friends and family to remind us to put our phones away whenever we pull them from our pockets at inopportune times.

Peter Bregman, author of *18 Minutes: Find Your Focus, Master Distraction, and Get the Right Things Done*, likens these sorts of interventions to tying oneself to the mast, as Odysseus did. The legendary Greek soldier was on his way home after ten years of war. He knew that, having pushed his mind and body to the limit, he would be unable to resist the song of the Sirens he would encounter en route to his wife and family. He wisely chose to tether himself to the mast of his ship, preventing himself from jumping into the arms of temptation. "It's so much more effective to create an environment that predisposes you to do the things you do want to do or don't want to do than to use willpower," as Bregman puts it.[2]

Use your calendar to plan monotasking. Putting to-do items in your calendar and planning dedicated time for nondistracted work offers two benefits. First, you are setting a clear goal and making a commitment to do the task at a specific time, increasing the probability that it will happen. Second, you are removing the task from your mental checklist, which decreases stress on your working memory. This frees up your prefrontal cortex for other tasks.

#2 Choose JOMO over FOMO

It is instantly rewarding to give in to the impulse to check our phones in response to every app ping and message ring. Many believe that a major driver of habitual notification checking and attention to social media is the fear of missing out, or FOMO. One recent study found a significant relationship between FOMO and problematic smartphone use, suggesting that higher rates of FOMO increase the probability of unhealthy habits related to social media. So turning FOMO into JOMO, the joy of missing out, can pay real dividends.[3]

Trading FOMO for JOMO means trading impulsivity for self-control. The long-term benefits of self-control are considerable; joy is worth the wait. But if it were easy to convince ourselves of this, and if it were easy to change the habit, there would be no FOMO.

FOMO isn't some newfangled feature of the digital age, although there are particular ways in which our broken relationship with technology fosters this particular anxiety. But really FOMO is a form of an old concept that scholars have thought about a great deal: delay discounting. Delay discounting involves lowering (discounting) the perceived value of a reward that is delayed, creating the tendency to prefer small, immediate rewards over large, delayed ones—for instance, the willingness to take a $50 payoff today as opposed to $100 next month. From an economic point of view, we are discounting the value of next month's promised payoff because waiting comes with some perceived risk. What if the promise doesn't come through? That uncertainty could be costly. The perceived risk makes the offer of $100 next month feel like $49 today—the discounted value—in which case, it seems sensible to go with the guaranteed $50 payoff now and forgo the possibility of more money later.

This conflict between short- and long-term payoff plays out across a range of human decision-making. We face frequent choices between today's easy reward and the greater long-term benefits from waiting. Your friend invites you to a party, but you have a big exam tomorrow. Do you go to the party or keep studying and go to bed early? Should you buy a new car, or save for your child's college fund and your retirement? You know you should exercise more, but you're hungry and tired. Do you go to the gym or eat a bag of chips and watch a movie? The list goes on and touches some of the most important personal and communal decisions we face—decisions that affect physical and mental health, including fitness and addiction, and social questions like whether and how to address climate change.[4]

When we face decision scenarios that pit today's instant rewards against tomorrow's delayed rewards, the discount we apply to the future benefit is a function of how much time separates it from the immediate benefit—time here is a proxy for uncertainty—as well as the value we assign to the two options. Choices therefore will vary across individuals. It is easy to imagine that the same person who would forgo the new car and instead invests in their child's education might also skip the gym in favor of chips and a movie. And while I might not agree with a student going to a party instead of studying, that choice would hardly surprise me. Discounting comes with an age effect: younger people, with their immature prefrontal cortex, are prone to higher rates of temporal discounting than are healthy adults. Adult decision-making isn't locked in, though; even among adults, temporal discounting is malleable under the right conditions. We can be persuaded, consciously and unconsciously.

Karolina Lempert is an experimental psychologist who studies decision neuroscience. I met her at a conference where she was discussing her work on the malleability of temporal discounting. I asked her two questions related to smartphone habits. First, could we frame the habit of obsessively checking our smartphones as a type of temporal discounting? In other words, are we trading the long-term rewards of greater focus, productivity, and intimacy for the immediate reward of checking our phones? The short answer is yes: smartphone habits fit the discounting model.

The second question: If temporal discounting is malleable, can people learn strategies for changing their smartphone habits? Again, the answer is yes, but with a caveat. As Lempert explained, the types of rewards our smartphones provide—such as likes and news updates—differ in important ways from those that have been studied in laboratory settings. When you ask a study participant to forgo $10 today in exchange for receiving $20 next week, that person will take a moment to think about their options. By contrast, our smartphone behaviors have become automatic. Smartphones provide very strong triggers, to which we often respond without any conscious consideration. Push notifications and the structure of social media so effectively take advantage of our bias for short-term rewards that choosing the better long-term reward is extremely difficult—more difficult than making the better choice of long-term rewards in an artificial lab environment. Fair enough. We knew changing habits would be hard. Still, "any strategy that would make people future-oriented or more mindful should work for both types of decisions," Lempert told me.

One of the key strategies Lempert has studied is the exploitation of framing effects. Framing refers to the way in which a decision is presented to the decision-maker. Marketers, lawyers, and sales professionals frequently use framing effects to influence consumers, jurors, and potential clients, respectively. For example, imagine a jury must decide whether a defendant is guilty of causing a deadly automobile accident. The key evidence is a security-camera video showing the incident. One lawyer might ask the jury, "How fast do you think the cars were going when they contacted each other?" Another lawyer might ask, "How fast do you think the cars were going when they smashed into each other?" People will tend to believe the speed is higher when the question is framed in the second way, even though the facts of the case are unchanged. Likewise, marketers know they should say, "Our product is 80 percent effective at improving your appearance," not "Our product fails to improve your appearance two out of ten times."[5]

One way to bias decision-makers toward the long-term good is to frame the decision as between a very small benefit today and a much larger benefit later. That is, the decision-maker should know not only that the long-term reward is considerably greater than the short-term reward but also that the benefits of the short-term reward are truly minimal. A person with a bad social media habit should know that reducing wasted time and distractions will improve their ability to focus, foster better work performance, and secure stronger relationships—all hugely consequential outcomes. But they should also know that setting the phone down and ignoring social media notifications means forgoing virtually nothing of importance. Being less up to date on inconsequential details of other people's lives is no great loss because this kind of knowledge does not foster greater well-being.

Relatedly, Lampert suggests framing the delayed reward with lots of detail about its value and benefits. This helps keep the longer-term goal top of mind, forces more consideration and deliberation about the future gain, and has been shown to reduce impulsivity. Are you striving toward a specific goal at school or work? Do you want to overcome the obstacles hampering a relationship with a friend, relative, or intimate partner? Imagine that goal in detail. Visualize a report card with an A grade, think about specific ways in which your life and responsibilities will change once you get that well-deserved promotion. Imagine the smiling face and sweet voice of someone that you care about saying, "Thank you" or "I love you."

Another factor to consider is our biased perception of time, which causes us to make choices today that our future selves would prefer not to make. Interestingly, heroin addicts have been shown to think in terms of very brief time horizons. Recall that the prefrontal cortex is key to our ability to visualize future states. People addicted to heroin, with their compromised brains, often see their future as the next few days, whereas healthy people will often see the next few months and years. This suggests that addiction can have a powerful impact on our perception of time and therefore influence temporal discounting dramatically, tipping the scale toward short-term rewards.[6]

The good news is that time perceptions can be modified by framing effects. The key is the human ability, aided by our prefrontal cortex, to use foresight. It turns out that people tend to be more patient when the future time period in question lies within the realm of imagination, and part of the way we achieve this is to imagine our future selves in detail. Taking the time to envision our future selves in detail keeps not only the future reward top of mind but also the future itself: we are more likely to accept a larger but delayed reward if our time horizon extends longer, so that a future event is readily imaginable.

The final framing device Lempert describes involves the emotional state of the decision-maker. We have discussed how negative emotions and stress can influence our self-control and make us more prone to unhealthy habits or addictive behaviors. Negative emotions such as fear also powerfully motivate temporal discounting, biasing us toward short-term rewards. Conversely, positive states, such as gratitude and appreciation, have been shown to make people more patient with delayed monetary rewards. Thus, to prevent destructive temporal discounting, we need to learn to handle stress and negative emotions in healthy ways. One suggestion is to be conscious about choosing JOMO over FOMO. When we realize that we are experiencing the fear of missing out, we have the opportunity to reframe fear as joy. Telling yourself that there are benefits to missing out is no guarantee that you will make the right decision, but the idea here is to be mindful enough of our mental states that we can intercede and thereby give ourselves the chance to make better choices.[7]

Psychologists, neuroscientists, and behavioral economists have made a great deal of progress in understanding the neurobiology of temporal discounting and decision-making. By now you won't be surprised to hear that neuroimaging and other research has consistently shown an important role for the prefrontal cortex. Studies show higher rates of delay discounting—that is, more impulsive choices—among individuals with drug or alcohol addiction, ADHD, conduct disorder, antisocial personality disorder, and borderline personality disorder, all of which involve the prefrontal cortex. Indeed, brain imaging has shown that the density of connections between the prefrontal cortex and networks involved in temporal discounting can predict rates of delay discounting. These are preliminary findings, but also exciting ones. It is possible that, at some point, we will be able to figure out which individuals are at greater risk of making poor temporal-discounting choices using only our understanding of the brain.[8]

Finally, we can be encouraged by the knowledge that delay discounting tendencies are not set in stone: even an impulsive person can learn to trade out FOMO for JOMO. The brain's neuroplasticity and neurogenesis enable rewiring—for better and worse—at any age. While prevention and early intervention are always the most expedient ways to manage unhealthy habits, thanks to our ability to rewire our brains, it is also possible to replace unhealthy habits with healthy ones. If one of the intervention strategies outlined above is not working, try another. There are lots to work with: reframing the perceived amount of time until a future reward arrives, visualizing the goal, describing future rewards in detail, accentuating the value of the future reward, and managing negative emotions and stress. All of these strategies can help people make better choices, including the choice to ignore the siren song of the phone or tablet and opt instead for future benefits, even when it feels better right now to get lost online.

#3 Manage Your Social Identity Carefully

In Shakespeare's *As You Like It,* the heroine Rosalind fears persecution by her uncle and flees to a nearby forest disguised as a young man. She later uses the disguise to teach Orlando, the object of her affection, how to be more loving and attentive. Unaware of Rosalind's true identity, Orlando listens and learns while being deceived. In her journey, Rosalind encounters a foil, who shares what has become a famous insight: "All the world is a stage," he says, "and all the men and women merely players."

More than three centuries later, in 1956, the sociologist Erving Goffman used Shakespeare's ideas to introduce his social-identity theory. In *The Presentation of Self in Everyday Life,* a book that would become hugely influential, Goffman draws on the imagery and setting of a Shakespearean theater as a metaphor for how humans present their self-identities in social interactions. On the frontstage, we create an image of ourselves, like actors performing for an audience. On the frontstage of daily life, the goal may be to make a good first impression or maintain social status at school, work, or in the community. This is a cultivated presentation of the self; we are acting, as if in a play, to manage the perceptions others. Backstage, we can break character and be ourselves, while a team of players—often close family and friends—supports our true identity, sometimes through reinforcement and sometimes through corrective action.

While Goffman could not have predicted that digital technology would be used to create entirely new stages online, with broader audiences and more players than we could imagine, he did foresee sources of trouble applicable to the online world. In particular, he suggests that social cohesion breaks down when our stories contain too many conflicts. This typically occurs when our external presentation—our frontstage identity—is too much in conflict with our private, backstage identity. As contradictions mount, we are likely to drop character while on the frontstage. Further complication arises when our team of friends, family, and peers is not aligned with our goals or when our front- and backstage goals become confused.[9]

The digital age brings a near-constant barrage of messages about friends, peers, family, politics, and news (fake and real) through a small screen and the many stages to which it provides access. We can extend our social identities onto a multitude of platforms including social media, multiplayer video games, and blogs. The audience provides instant, quantifiable feedback in the form of likes and comments on our performance. Amid this ferment, our performed identities multiply, more easily coming into conflict with our backstage selves and generating more challenges for our team to cope with—the sorts of challenges outlined in part 2.

Am I the person I present on assorted digital stages or on the frontstage of real life? Is my team composed of those players I interact with in the real world or in the world of video games and social media? Eventually Dunbar's number will impose limits, forcing me to choose. And is there a backstage anymore, in a digital world where nothing is private and many of us share our every move?

The ease with which we curate and transform our self-identities and guide our audience online creates as much potential opportunity as it does peril. The boundaries between various frontstages become overly complex, and the many identities we cultivate become difficult for us and our team players to manage backstage. The more effort we need to put into our capacity as performers on the various frontstages, the more of our lives we will spend at a distance from our true, backstage selves. The widening gap between our front- and backstage identities has the potential to further stress our prefrontal cortex, which works to coordinate and conduct our various selves and associated emotions online and off. Managing multiple social identities contributes to stress, strains real-world relationships, and increases the risk for mental health problems including anxiety, depression, loneliness, addiction, and suicide.[10]

Thus the third recommendation for maintaining tech-life balance: be mindful of which versions of yourself you present online and of the reasons you choose these self-presentations. The more identities we create and the more frontstages we try to manage, the harder it will be to manage our true selves and our team players. We live best when we are able to gain support from the people who matter most to us, but sometimes our online lives put us out of their reach or challenge ourselves and our teammates in ways we are not prepared for.

The difficulty here is especially great for tweens and teens who are in the most formative stages of crafting their identities and who face the greatest pressure to take part in the various stages of the online world. Recently, I was spending some time with a friend and her eighteen-year-old daughter, who divulged to us that she had shared a photo of herself in a bikini on Instagram. The post quickly accrued over a thousand likes. As she smiled at the near-instant feedback, I asked her if all her photos were so well received. No, she explained, only the ones of herself alone in a bikini. Her mother was mortified. It never occurred to this teenage girl that all the attention her half-naked body receives online could be a burden to her family and to herself. Friends, parents, and health professionals can and should remind impressionable young people of the complexities of exposing too much online, because our many identities can haunt us now and in the future.

Adults, too, can manage only so many social identities, and we must be careful with the details of each. Experimenting with our identities on various online and offline stages is inevitable. But when using social platforms, we need to consider the differences between public and private sharing. We need to think about whether groups of "friends" or people who follow us are truly team players on our life's stage or instead part of our audience. Social platforms can and will be part of all of our evolving social identities, but they demand care. We must take our online behaviors and interactions seriously, just as we do their offline equivalents. This leads us to the fourth recommendation.

#4 Think before You Post

Roseanne Barr was enjoying an epic television comeback. After a successful run as the lead actress in her eponymous show in the 1990s, she returned to ABC in March 2018 with an updated program in a prime-time slot. After nearly twenty years, the revival was a hit: *Roseanne* garnered over 16 million viewers, making it one of the most-watched new television shows in years. The network quickly signed Barr and the cast for a second season and made the show a centerpiece in its annual pitch to advertisers.[11]

But by late May, *Roseanne* had been canceled. Everything came crashing down after Barr posted a racist tweet about Valerie Jarrett, an African American former aide to President Barack Obama. If the "muslim brotherhood & planet of the apes had a baby = vj," Barr wrote. She immediately tried to offer an apology, describing the comment as a "bad joke" and "in bad taste." But the response from ABC and its parent company, Disney, was swift and final. Barr was fired and a month later was quoted in the news media saying, "I lost everything." Her failure to consider carefully the consequences of what she was about to say on social media altered the course of her life—forever.[12]

If our personal identities are socially constructed, influenced by the assorted frontstages on which we present ourselves, then what and how we communicate through social media and other digital platforms is critically important. The frontstage of social media can be very powerful, amplifying messages as our connections grow and enticing us with constant feedback. The Barr example is extreme in its content and the response it received, but it is a reminder to everyone to think before they post.

By everyone, I do mean everyone—not just celebrities, public officials, and others who are more likely to incur backlash. Backlash generates headlines, but the fundamental evil of intemperate, foolish, and bullying behavior online is that it hurts others. The problem is epidemic in schools, which face the challenge of policing behavior not just on campus but also on social media. One school district, the Mounds View Public Schools just north of Minneapolis, created a handy mnemonic to urge students to review their online comments before sending them into the world. It's a simple riff on a phrase for our times: "T.H.I.N.K. before you post." Is what you're posting True? Is it Helpful to someone or some group? Does the post Inspire? Is it Necessary? And is it Kind? The school district's campaign includes examples of social media posts gone wrong, such as a high school lacrosse player who maligns his coach for not playing him in a game and a basketball player taunting a teammate for playing poorly in a loss.[13]

Before posting, ask yourself some questions. Would you ever speak this way offline, knowing that lots of people would hear you? If the answer is no, then that comment is probably out of bounds. The empathy you feel offline should not simply evaporate online. After all, you are not shielded from shame by the digital distance of social media platforms. Nor do others cease to be vulnerable just because they are absorbing your comments from behind a screen. And how would you feel if your parents, friends, loved ones, or employers—or, indeed, your future self—were to someday see what you are posting today?

Like bullying offline, mean-spirited online posts reflect the hardships of the person responsible. Venting online can be a strategy for managing negative emotions, albeit a poor one that is unlikely to result in effective and lasting solutions. And it hurts people along the way. This is an instance of one of our key themes: acting impulsively to achieve catharsis, even as one sacrifices the long-term benefits for oneself and others.

The Mounds View T.H.I.N.K. mnemonic is a good one for young people because it asks them simply to consider actions and consequences before they click. Adults, however, should be mature enough to reflect on their own states of mind—the sources of their actions. What is causing you to behave inappropriately online? Perhaps you are getting caught up in social comparisons and feeling insecure and inadequate. Maybe you are incensed by misinformation. If you are using social media as a mood regulator, then what is the source of your foul moods? Our behavior manifests the feelings we are experiencing; that is the case offline and on.

We need to create space to have honest conversations with ourselves. To that end, we also need strategies to stem problematic behaviors in the moment. Because offensive online behavior is often impulsive, several of the strategies discussed with respect to delay discounting apply here. Project yourself into the future and imagine more important goals. Imagine those goals in detail so that you can feel the weight of their benefits. These strategies can slow us down, preventing us from harming others and forcing us to confront the underlying causes of our behavior.

We know that there is a lot of good to be derived from social media. Communication tools can help us strengthen our bonds with friends whether long-standing or new. When used appropriately, social media can amplify feelings of happiness and contribute to our social lives offline in the long run. Social media is an ideal vehicle for sharing news with distant relatives and friends and for scheduling events that strengthen real-world social bonds. Like any tool, however, social media can be misused—and misuse is ever so easy when the platforms are designed to put contentious arguments and provocative posts at the top of our feeds. We need to recognize, though, that these algorithmic outcomes, too, reflect our collective states of mind. If Facebook is showing you lots of posts that anger or depress you, spur you to outrage, or leave you feeling envious and insecure, that's at least in part because you've engaged with such posts in the past. These are signs that you are putting yourself in an unhealthy position online.

Think before you post, and think before you become involved in arguments, insulting threads, and social comparisons others instigate. Use the blocking functions built into most platforms to ensure that you don't see posts from instigators. If necessary, resolve to post only positive messages and verifiable facts. Review your posts to assess whether you're meeting your goals in life, which should include staying close to family members and real friends.

#5 Prioritize Strong Social Bonds

The ability to form strong social bonds is a key factor in humans' unique success on this planet. As discussed in part 1, social bonds are critical to the developing brains of our children. Social bonds also helped us become efficient hunters and gatherers and later farmers and settlers who learned to work in cohesive groups that sacrificed individual and family needs for the greater good. Strong social bonds allowed us to organize complex societies, share information, and collaborate to advance technology and cure diseases.

Today, there are many challenges to forming and maintaining strong social bonds, from the distractions of every app ping and message ring to the divided, automatic attention induced by the mere presence of our smartphones to the 24/7 always-on work culture. At the same time, levels of stress, anxiety, and depression are sky high. Social bonds help us handle stress, yet at precisely the time when we need social support the most, we are succumbing to technology habits that undermine it.

We need to prioritize strong social bonds and intimate relationships. If we don't put time and effort into these relationships, they will deteriorate. Given this need and our limited capacity, we must be thoughtful about how we nurture healthy relationships. This means we need to be clear about the differences between the deep, intimate, and authentic connections we make offline and our shallower and more curated social connections online.

As we have seen, our health and happiness are dependent on appreciating this distinction. In fact, strong social bonds may be the difference between life and death. A bold statement, but one with empirical support. A pioneering 1988 paper introduced this idea on the basis of anecdotal evidence suggesting a link between social relationships and death rates. The authors collected scores of case studies concerning married couples, in which one partner would die and the other would follow a few months later. This raised the question of whether the loss of a supportive life partner could have direct consequences on the surviving partner's health, contributing to their untimely demise.

Inspired by this idea, researchers began looking more rigorously at associations between life expectancy and the quantity and quality of social relationships. By 2010 a group of researchers were able combine 148 studies in a meta-analysis involving over 300,000 participants in North America and Western Europe, subjecting this provocative hypothesis—that better social relationships make for longer lives—to thorough scientific scrutiny.[14]

The results of the meta-analysis unambiguously support the hypothesis: social relationships were a significant factor in life expectancy, conferring a 50 percent increase in the odds of survival. Put differently, this means that those without adequate social support are 50 percent more likely to die prematurely. The absence of significant relationships was a more important risk factor for premature death than physical inactivity and obesity and was comparable to chronic cigarette smoking and heavy alcohol use. The results stood up over time and across countries. They remained consistent regardless of age, gender, initial health status, cause of death, and other potentially relevant factors.

One of the limitations of the review was that many of the studies assessed only whether individuals were married or lived with others. Yet it is possible for someone to live alone and have a healthy social network, while someone else could be married but also isolated and lonely. Further analyses have taken up more complex measures of social bonds—metrics that gauge the number and strength of those bonds—and found an even tighter link between our social life and life expectancy. The precise underlying mechanisms are not clear, and probably there is variation in individual cases. Social relationships can be valuable for mental and physical health, as friends and loved ones can help us relieve stress, encourage us to exercise, and take us to doctor's appointments when necessary, for example. What we know with confidence is that the more robust and reliable are our social networks, the greater the benefit to our health and longevity.[15]

Longevity is not the only benefit of strong social bonds. Let's return to Bob Waldinger, the psychiatrist and director of the long-running Harvard Study of Adult Development. One aspect of the study focuses on what makes us happy in the long run. Waldinger starts his enormously popular TED Talk—40 million views and counting—with a simple question: "If you were going to invest now in your future and best self, where would you put your time and your energy?" He then cites a survey of US Millennials that shows their most commonly reported life goal is to be famous and get rich.[16]

But is this what really makes us happy? Waldinger's decades-long study suggests otherwise. The results show that wealth, fame, and even hard work do not appear to correlate with healthy outcomes and do not make us happier. Rather, "good relationships keep us happier and healthier, period." From this we can infer a sobering insight concerning our social media relationships: they aren't real relationships. At least, they aren't relationships of high quality. As we saw, research has shown that social media use produces no overall increase in well-being, yet good relationships lead to both longer, healthier lives and increased happiness. Ergo, social media should not typically be understood as bringing people together in any meaningful way, no matter what Silicon Valley marketers suggest.

Social media is great for fostering lots and lots of superficial connections, but superficial connections don't do us much good. Recall Dunbar's theory, which holds that there is a limit—quite a low one—on the number of truly valuable relationships one can have. Our brains are built to support about 150 substantive relationships and only a handful of intimate ones. Our close partners affect us deeply, for good and for ill. Perhaps not surprisingly, Waldinger has found that high-conflict marriages with low affection are almost as bad as divorce in terms of health consequences. Meanwhile, what predicts positive outcomes is satisfying social bonds with just a few individuals. These are the relationships that have the greatest positive impact on life expectancy and overall well-being, including brain function late in life. Researchers have found that those with strong social relationships have better memory capacity and less dementia as they grow older.[17]

The question for our digital age is how we can rescue close personal relationships from the distractions, social comparisons, and head-fakes of social media. Again, most social media relationships are not actually relationships but digital "friends." But when used judiciously, social media can enhance true relationships. We must be careful to use social media to support bonds that matter.

To this end, think about how to transition social media connections into face-to-face social interactions. If your connection is far away, talk on the phone or use video-chat. As we saw, video-chat is not an adequate substitute for in-person presence, but the synchronous timing fosters more of the emotional connection we need than does asynchronous, text-based communication. Better still, schedule a social event where you can meet in person, share an experience, and exchange stories.

Another option is to use media as a social event in and of itself. Appointment television may be a thing of the past, but you can still invite friends over rather than bingeing Netflix yourself. And parents should make an effort to be present while their children are consuming media. As we saw, kids learn more from educational videos when a parent or supportive caregiver reinforces the lessons on screen. Studies even show that playing violent video games with friends has different consequences than playing alone. The social aspect of video game play appears to decrease the risk that violent gaming will lead to aggressive behavior.[18]

We also would be well served by cutting off social media connections that undermine our emotional well-being, making us less empathetic and therefore less able to form and maintain strong social bonds. If your online time is spent with social comparisons, bullies, and narcissists, that is time poorly spent. When we come upon the rants and diminishing insults of online "friends" and faceless followers, we do well to reframe these experiences not as criticisms of ourselves but as evidence of other people's struggles. Those struggles do not have to affect us negatively. Indeed, they can be opportunities to practice empathy for people who are lashing out and lost in their own cyber-reality—people who may be suffering or insecure.

And let there be no doubt that empathy can be practiced and learned. Helen Riess, a psychiatrist, friend, and past collaborator of mine, has worked with physicians late in their training when "compassion fatigue" and burnout set in, leading to reduced empathy with patients and others in their lives. She found that physicians who received just three hours of empathy training earned higher empathy ratings from patients than did a control group of physicians who did not receive the training—they instead saw their empathy ratings fall. The three-hour-long course includes an overview of the neurobiology of empathy and training in specific skills, such as interpreting nonverbal communications from patients. The course is now available to all health care providers online and through an app—an example of how technology can be used in healthy ways to improve ourselves.[19]

Children, too, can improve their empathy. Researchers at UCLA recruited a group of public school sixth graders to attend an outdoor camp 70 miles outside of Los Angeles where they would have no access to media technologies for five days. Compared to a control group of sixth graders from the same school who did not attend the camp, those who took a break from social media and spent nondistracted time at the camp showed significant improvements in measures of nonverbal emotional understanding—a key component of empathy. Both the physician and student studies are hopeful signs of the brain's resilience. Under the right conditions, the brain uses and builds specialized networks that enable communication and mutual understanding. We are wired for both, and we can be rewired for the better.[20]

The effort to improve our social relationships, like the effort to think before we post, calls on the important skill of delaying gratification. The apps and devices that call for our attention, distracting us from others in the room, provide immediate rewards at the cost of deepening relationships that matter far more for our health and happiness. The imaginative strategies associated with goal-setting are important here, too, so we can train ourselves to resist the impulse to check every app ping and message ring. And when you are with others, make sure your smartphone isn't visible. Now and then, technology contributes to the conversation, such as when looking for travel directions or connecting with others in a group to join later. But often it is simply a distraction that has the high potential to disrupt the relationship with the person in front of you. And because our phones bother us even when out of sight, use the do-not-disturb function to silence the ringer and arrest vibrations when socializing.

This is not a suggestion intended to ensure constant tranquility. Most of us are not monks, and sometimes urgent messages do arrive while we are in fellowship with others. In these cases, don't go back and forth texting, an inefficient and time-consuming method of communication that drags out the interruption. Apologize to those around you, go to a discreet place, and make a phone call. Resolve the issue as much as possible and then rejoin the group and apologize, again. This type of digital etiquette acknowledges the disruption, models good behavior with technology, and turns a moment that might fray social bonds into an opportunity to nurture them.

#6 Don't Fall for Compulsion Loops and Clickbait

A sexy celebrity has the secret to a thin waistline. An impossibly young-looking older man has a new pill for longevity. A smiling model promises a fast track to financial freedom. These provocative images are superstimuli, and when combined with a tempting caption, they can be irresistible. Clickbait captures our attention and gives us just enough information to arouse curiosity. At best, we are distracted for a moment. At worst, we fall into the trap.

One particularly concerning form of clickbait is so-called food porn. Imagine a photo of rich vanilla ice cream, melting ever so slightly so that just the right amount drips over a sumptuous mound of warm peach cobbler. It is hard to resist. There is even a website called FoodPornDaily, with the motto "Click, drool, repeat." That tagline perfectly sums up the goals of clickbait and compulsion loops—the marketing tools that online content creators have perfected in order to draw you in and keep you coming back for more. And it works. When we are hungry and even when we are not, food is highly salient and relevant. We don't have to hide food porn either, as it is socially acceptable.

It is also a growing problem. Despite widespread warnings concerning the health effects of weight gain, the epidemic of obesity continues. Estimates suggest that if current trends continue, nearly half of all Americans will be obese by 2030. Obviously food porn is only one inducement to overeating; we saw, for instance, that excess media consumption and exposure to food advertising tricks the brain's satiety centers and leads to excess calorie consumption in children and adults alike. But food porn's sudden online proliferation is worrying, in part because its aesthetic is also especially powerful. Food marketing has historically been an area full of superstimuli—think of the old cereal box disclaimer "Enlarged to show texture," which really means "Enlarged to induce hunger." Food porn takes this approach to a new level, deliberately accentuating the least healthy foods in particular, in order to catch our eye. Eye-tracking studies clearly show the human visual system prefers foods that we associate with high fat content. Images of such foods trigger the brain's reward centers in ways that low-fat foods don't.[21]

Even more intensively than traditional food marketing, food porn hacks our brains, taking advantage of its evolved capacities. The brain is wired for visual hunger—the instinct to look for and at food. Our brains evolved to create reward responses to the sight of food, since seeing food correlates with eating it, and that, to a point, is critical for our survival. But for most of us, food is not as scarce as it was in the early days of human evolution and throughout much of the history of our species. When food is plentiful, visual hunger is no longer an evolutionary adaptation, but images of food can still tempt us. Food porn puts our hunger instincts into overdrive, contributing to weight gain.[22]

Most insidiously, our brain wiring for visual hunger, combined with easy food access and ubiquitous supernormal food imagery continuously displayed through social media, sets the table for people predisposed to obesity. Obese people have higher anticipatory responses to food, meaning their brain response to the mere thought of food triggers higher neural activity as they anticipate an expected reward. In concrete terms, obese people experience stronger food cravings than people with healthy weight. In addition, compared to people with healthy weight, obese people experience less activity related to the sensory pleasure of eating. That is, obese people have a tolerance to food. In both of these respects, obesity is much like an addiction: the brain, having adapted to a stimulus, gains greater rewards from the anticipation of the stimulus than from the experience of it. Yet while society would never tolerate "heroin porn," and over the course of decades became sensitized to the dangers of glorifying cigarettes and nicotine addiction, food porn suddenly is everywhere, creating hazards for those who may be facing the greatest difficulty with their weight.[23]

Visual hunger is a form of automatic attention. This is the nature of all clickbait: it preys on our unconscious tendencies to attend to whatever is salient in our environment. Initially, only conscious inhibition of that attention can direct us away from the stimulus attracting it. However, over time, we can build the habit of recognizing the features of clickbait: position on the page, overly attractive people, offers that are too good to be true. Once we have a feel for what clickbait looks like, we can more readily ignore it without having to consciously evaluate its characteristics. With respect to food porn specifically, research suggests that exercising and getting a good night's sleep can help us avoid the temptations fostered by hunger-inducing images.[24]

Unfortunately, avoiding clickbait is not as simple as recognizing classic bogus ad formats. This sort of clickbait has been around since the early internet, and at this point is so well known that it's the subject of parody. But while we've learned to make fun of clickbait, content creators have developed more sophisticated ways to catch our eye and keep us clicking. Food porn is just one of the more effective manipulations. Creators have also developed engaging features and applications: video games that withhold game-play features early on, pushing us to invest more and more time in them; autoplay video that keeps us watching for hours; and the infinity scroll, which puts an endless number of social updates and news items in our feeds, providing users a bottomless pit of intrigue and potential rewards.

Basically, the goal of software developers has changed. The idea is not so much to create useful tools or platforms for learning specific pieces of information. Instead, across nearly every corner of the internet, nearly every kind of application is designed to integrate compulsion loops in order to foster new habits. The only tool some developers make at this point is a hook—and you are the fish they are trying to catch. This is especially true of free applications. Profit comes from advertising, which is easier to sell when there are many users with high engagement. And user engagement is highest when users engage habitually. So developers, informed by some of the same brain and behavioral science described throughout this book, purposely create sites and app features that stimulate our reward centers, tricking our brains to create automatic actions that keep us stuck in the loop.

The classic compulsion loop is designed to either relieve some stress or trigger some reward, or both. Almost all video games are based on compulsion loops. The player becomes aroused and tense while negotiating the game-play challenge and, upon completing it, is rewarded with positive reinforcement—points, a win, or advancement through a "level-up." The tension is thus momentarily relieved and replaced with gratification. But then, more or less instantly, another opportunity to repeat the process is made available. Owing to computational, memory, and platform limitations, older games eventually came to an end and had other natural breaks; the cycle could repeat only so many times. Today the cycle repeats endlessly.

Social media embeds its own compulsion loops, some of which resemble those of games and the gamification of many actions. The backbone of social media is the compulsion loop created by new updates. People post updates of varying interest, creating a highly variable reinforcement schedule. This keeps us coming back for more: we're never sure when our engagement will be rewarded by an interesting post, so we keep checking. Push notifications—those app pings and message rings—exacerbate the situation, telling us to check *now*. This is also how social media functions as a mood regulator. Whether we seek relief from stress or from boredom, social media provides through the promise of endless updates.

Compulsion loops foster repeated exposure, which in turn fosters new habits. This process takes time and, at first, conscious choices. This is the exploration phase, when we first learn about some tool or stimulus and begin understanding how it works. Perhaps a friend introduces us to an app or social media feature. After initial trials, we experience some sense of relief or enjoyment; this is the reward the app or feature offers. In this early period, we learn to associate a need discovered through exploration with the reward produced by fulfilling that need. Of course, this is not a real need; it's a sensation created to hook the user.

In time, this association between need and reward becomes firmer through repetition, and eventually, as we learned, the striatum takes over. Induced by the continuing flow of rewards, the striatum creates an action unit so that we don't even need to think about what we are doing. This is the point where conscious action gives way to habit. The final stage is looking forward to the next experience—anticipation. The compulsion loop—stimulus-reward, stimulus-reward, ad nauseam—eventually breaks down the self-awareness and self-monitoring performed by our prefrontal cortex.

The brain changes associated with compulsion loops are clearest among heavy video gamers. A 2014 review of more than twenty brain-imaging studies demonstrates unambiguously that the brain activity among heavy gamers is consistent with the brain activity associated with cravings, reward response, and changes in the prefrontal cortex seen in patients struggling with addiction. In time, we will likely see similar findings with heavy social media users.[25]

Defeating the habits and addictions fostered by online compulsion loops is like beating any habit or addiction. It requires a plan, a commitment, and social support. Pick a date. Visualize a detailed future version of yourself living a healthy lifestyle with fulfilling offline relationships. Commit yourself to stopping the behavior. Seek professional help and the support of your friends and family. Expect setbacks and look to replace the existing behavior with a less destructive one—the aforementioned temptation bundling. These steps are the key to the experiential reboot that will bring your brain's reward system to the point where you can once again experience the natural rewards of life. This process can take a good deal of time, but it need not. Some of us have probably experienced this kind of detox. Maybe you recall when you were "addicted" to a simple online game like *Candy Crush* or *2048,* but you got over it. Younger people and heavier users of certain games, apps, and social media will need more time. But anyone can change their habits.

Of course, avoiding an unhealthy habit in the first place is easier than unlearning one. This is why the most successful approach is to recognize the superstimuli that drive clickbait and compulsion loops and proactively steer clear of them. Hopefully the discussion in this book has sensitized you to the issues and inspired new behaviors. Because mobile media, communications, and information technologies are useful to us, and because social media and games are fun, most of us will indulge now and then. This is not a problem. Problems arise when indulgence turns to habit (and then addiction) and takes over, leaving insufficient time and energy for what really matters—productivity in work and

school, fulfilling hobbies, independent learning, and building strong relationships. As long as we are aware of what we are doing and able to intercede before we become carried away, we can safely derive pleasure from social media,
shopping apps, and online games. We just have to be vigilant. When you start
up that app or game, tell yourself it is a diversion. Set a timer and log off when
it alerts you, because the app, game, feed, and video stream won't stop on their
own. Until the economics of the internet change radically, it will be up to you
to set a limit and sever the loop.

#7 Choose Paper over Pixels

In 2011 technology entrepreneur Jean-Louis Constanta posted to YouTube a provocative video of his one-year-old daughter. She happily plays with an iPad,
then turns to a paper magazine. She pinches the magazine, but nothing happens. Visibly perplexed and unhappy, she struggles with the printed page and
crinkles her face. Why won't the magazine respond to the touch of her finger?
The video, titled "A Magazine Is an iPad That Does Not Work," is a stark message and a testimony to the changing ways in which we learn and process information in the digital age.[26]

As we saw, reading is a skill acquired with concerted effort. That isn't about
to change, given the need to recruit and train multiple brain areas and capacities at an early age in order to read. But what if digital tools could make reading
easier? What if a magazine really is an iPad that doesn't work—an outdated
vehicle for disseminating written information? As computer-based education
has taken off in the classroom and at home, it is worth asking whether we might
read and learn as efficiently on a screen as we do on paper.

This is known as the paper-versus-pixel question, and it has been subject to
rigorous study. As of this writing, more than a hundred papers have addressed
the issue, incorporating perspectives and experiments from psychologists, neuroscientists, computer engineers, library scientists, and others. Like many of
the issues addressed in this book, the paper-versus-pixel question has not been
definitively settled in every respect. But there are telling findings. A review of
the existing evidence suggests that even Millennials and members of Generation Z, the digital natives, tend to process and remember information printed
on paper more effectively than they do information appearing in digital text
on a screen.

At this point the debate centers on why paper is a superior reading and learning medium. There are many theories. Some research suggests that reading on a screen is more mentally taxing. It can also be more physically taxing; recall the discussion of computer vision syndrome. Other theories propose that we bring to computer screens a mindset that courts distraction: when we look at screens, we expect easy access to a world of endless information and titillation, triggering our brains for continuous partial attention and multitasking rather than deep learning. A printed page, in contrast, fosters more sustained and focused attention.

Still other theories suggest that web browsers and e-readers, despite their improvements over the years, lack certain tactile features that cue us to our progress and help us navigate reading with greater ease and less distraction. The tactile experience of an object as an aid in our use of that object is known as haptic feedback. Physically turning the printed page, the shifting weight of a book as we progress through it, inserting a bookmark or dog-earing a page— these are all physical cues that guide us through reading. Haptic feedback is often missing or less salient with digital reading, which lacks the physicality of the printed page.[27]

And while e-readers offer a less distracting environment than most websites, many incorporate hyperlink text intended to inform, which has been shown to create disruptions and reduce comprehension. Visually, e-readers are generally quite good, but they still lack certain navigational cues and progress markers. For instance, one of the conveniences of e-readers—their minimal thickness—is also an obstacle to gauging progress. The thickness of a book, growing on the front end and shrinking on the back end as we read, clearly indicates progress.[28]

There are also a number of ways in which print improves reading comprehension. Page-turning allows a brief but natural break for reflection and memory consolidation, which we sacrifice with the instant flip of the e-reader and the digital scroll of the web browser. The experience of reading a paper book, including page-turning, also has been shown to engage a broad set of brain networks that complement the visual system and language centers, inspiring a coordinated neural response that facilitates focus. It is as if the conductor that is the prefrontal cortex has additional instruments to make the symphony of reading richer and more harmonious and therefore easier to understand. Another aid to comprehension lies in our ability to recall passages by their physical location on a printed page. Research shows that e-readers and digital screens do not measure up on this score. Even advertising is better processed and remembered in print form on paper than in digital form on a screen.[29]

Finally, e-readers come with a barrier to entry that affects young kids' ability to benefit from them. As discussed, research shows that parents using e-readers and apps designed to tell their children stories with words and pictures spend more time commenting on which buttons to press and how to navigate the experience and less time discussing the characters, their emotional worlds, and elements of plot and setting. Children become engaged in the mechanics of the publication rather than constructing the story in their minds, depriving them of opportunities to create connections in a budding neural circuitry crucial for realizing their full reading and learning potential.

This is not to deny the advantages of e-readers, which allow us to carry a library everywhere and which can be filled with digital books that cost a fraction of paper ones. But e-readers may not be the best tools for learning to read and for comprehension. The recommendation here is simple. When reading for deeper understanding, such as when studying for an exam or learning new and complex material for work or self-fulfillment, print the relevant document or buy a bound book. Paper improves learning efficiency and reduces eye strain. It inspires monotasking and provides greater haptic feedback to help us focus. And it aids retention. Finally, since paper doesn't emit light, it is perfect for reading at night before you go to sleep. Which brings us to the next recommendation.

#8 Don't Bring Technology to Bed

We spend nearly one-third of our entire lives in a blinded, semiconscious state of almost complete paralysis. Nearly all species of animals do it, despite the obvious vulnerability such a condition creates in the wild. Why do we do it? Why do we sleep? It may surprise some readers to learn that scientists don't have a complete answer to this question. But all of us know that when we don't sleep enough, we feel lousy and are less capable and alert throughout the day. We can also be confident that if sleep were not absolutely necessary, we and other animals wouldn't subject ourselves to the risks it creates.

I don't want to oversell the point, though. While sleep remains something of a mystery, we are not entirely clueless about its importance. One thing all sleep scientists agree on is that the act is critical for memory function and therefore essential to learning. The question, then, is how sleep helps us with memory and learning tasks. Do we sleep to remember or sleep to forget? That is, does sleep help our brains consolidate the memories we acquired throughout the day for future use, or is sleep a state in which we let go of memories that serve little purpose, the better to make use of our limited memory capacity and to store information that is really important?

One leading theory, based on studies of humans and nonhuman animals, suggests that new information learned throughout the day is replayed at night. By replaying neuronal patterns as we sleep, the brain strengthens relevant neural circuits for future use. This is the sleep-to-remember hypothesis: we preferentially reinforce the connections and strengthen the synapses between neurons to make memories last—a process critical for learning.[30]

On the other side of the debate is the sleep-to-forget hypothesis. Researchers have found evidence that neuronal connections are weakened during sleep, reducing neuronal stress. This suggests that we forget certain information while sleeping and thereby replenish our cognitive abilities for the next day. The authors of one important study speculate that as we sleep, our brains experience a "neural free-for-all" that features spontaneous activity across many brain networks and weakens useless memory traces that clutter the brain. It could be that this neural free-for-all accounts for the randomness in many of our dreams.[31]

As research has progressed, we have learned that sleep is important not only for memory but also for enabling problem-solving and creativity. In particular, REM sleep has been shown to enhance the integration of information in the service of these goals. The stages of sleep always unfold in sequence, and REM comes relatively late, so we need a good night of sleep in order to achieve full restorative benefits. And, as everyone knows, the alternative is not healthy. We saw how reduced sleep has been linked to a higher risk of automobile accidents, substance abuse, poor school performance, obesity, and mental health problems. Reduced sleep has also been linked to higher rates of Alzheimer's disease. And we saw that more and more people across the world and in every age group are not getting enough sleep, exposing us to these serious hazards.[32]

Undoubtedly numerous factors underlie reduced sleep, but one of them is almost certainly the widespread use of digital screens late at night and during periods of wakefulness when we should be sleeping. Surveys from the US National Sleep Foundation show that the majority of Americans use some kind of digital screen or watch television in the hour before sleep at least a few nights a week. And we saw that the overwhelming majority of American adults and kids have at least one electronic device in their bedrooms. Large percentages also send text messages, use social media, and read news online when we periodically awake from sleep during the night.[33]

The reason this is a problem is that our phones, tablets, laptops, and most newer televisions emit predominately blue light. (This doesn't mean that we perceive all the colors on screen as blue, but the light emitted by our screens is skewed toward the short-wavelength, or blue, end of the visual spectrum.) Blue light disrupts sleep and reduces sleep pressure because it is also the kind of light emitted during the day and by the many artificial light sources we have created. Over the course of evolutionary history, human and most animal optical-processing systems have learned to equate blue light with daytime. Specialized receptors in the eyes detect blue light. This signal is transmitted to the brain via the optic nerve, and the brain directs the adrenal glands in our kidneys to produce more cortisol—the hormone that puts us on alert and prepares us for action. In the evening, the predominantly longer-wavelength light of orange and red sunsets sends a different signal to the brain, triggering the slow rise of the hormone melatonin, which signals the brain that it is time for sleep. This daily cycle plays out below our consciousness, driven by Earth's spinning axis and relationship to the sun. This cycle has influenced the timing of sleep and work since the dawn of humanity.

The LED (light-emitting diode) screens in our devices are bright and blue, and they have become brighter and bluer over time, contributing to sleep problems. The harmful impact of screens on sleep is by now so widely known and so widely felt that engineers and entrepreneurs are creating products to counteract the problem. These range from specially tinted glasses that filter out blue light to software that adjusts the color spectrum of a screen's output. It remains to be seen, however, whether these solutions are effective. Early studies suggest that new devices and technologies can reduce blue light from screens but cannot eliminate it, and they do not reduce eye strain. Plus, even if a screen is programmed to emit orange light, users will still be engaged with the content on their devices—content designed to capture our attention, trigger emotional responses, and keep us alert.[34]

Ultimately, the recommendation here is simple: practice what experts call good sleep hygiene. Don't watch television in bed, and leave smartphones outside of the bedroom in silent mode overnight. Buy a cheap alarm clock so you don't need your phone alarm to wake up. At least one hour before bedtime, but preferably two, put all electronic devices away, turn off the television, and turn down the lights to send a signal to your brain to naturally elevate melatonin levels. At this point, taking a time-release melatonin supplement can also be helpful. Then, read paper books and magazines or try to meditate. Keep regular hours; go to bed and wake up at the same time every day. Some of us may need prescription medications to overcome insomnia, but these can alter sleep architecture and often have side effects. They should be a last resort, when healthy sleep hygiene alone is not enough. And parents need to practice good sleep habits on behalf of their kids: put away their electronic devices at least an hour before sleep, set clear rules about media-technology use at night, enforce consistent bedtime hours, and be mindful throughout the day of children's intake of stimulants such as caffeine.

#9 Put Your Phone Down While Driving

In most cities, automobiles began as a luxury for the rich and a curiosity for the rest. But Detroit was different. As the home of Henry Ford and his soon-to-be-famous company, Detroit was a testbed for the mass use of affordable cars, far outpacing other cities. Ford's Model T, released in 1908, eventually revolutionized travel everywhere, but Detroit was first.

While automotive technology delivered on the grand promise of improving mobility, in the early days it was also incredibly dangerous. In Detroit, automobiles suddenly appeared in droves amid the jumble of streetcars, horses, carriages, peddlers, pedestrians, and children at play. People knew how to handle streetcars and horses, but the cars were unfamiliar and unregulated, and there were no rules of the road. Unaccustomed to the speed and braking distance of cars, many people found themselves in harm's way. Few doubted the promise of automobiles, but frequent collisions and the resultant injuries and deaths of adults and children were a source of anguish and frustration.

It took years for the public to adopt meaningful rules that responded to the growing number of cars. Downtown Detroit got its first stop sign in 1915, seven years after Model Ts started creating chaos on the roads. The first patent for an automated traffic signal using red and green lights to control the flow of vehicles was issued two years later. Over time, more rules were established to make driving safer and ensure that people were prepared for the task. States mandated licensing, driver education, and insurance and created laws against driving under the influence of alcohol and drugs. Other laws mandated seat belts, brake lights, windshield wipers, and eventually airbags, emissions standards, and annual inspections. New rules more effectively governed the flow of traffic on roads that were made safer through paving, widening, and the installation of reflectors and streetlights. The new discipline of graphic design was recruited to create more visible signs that clearly and quickly conveyed the information motorists needed to drive safely. In the late-twentieth century, automobile reliability improved dramatically, further reducing hazards on the roadways.[35]

For over a hundred years, we've wrestled with auto safety, and for most of that time, things got better. But that is no longer the case. In 2016 the National Safety Council, a US nonprofit, reported that after decades of declining deaths on the highways, the country was experiencing a surge in traffic fatalities—up nearly 6 percent in one year. It was the largest increase in over fifty years. In 2018 the death rate surpassed 40,000 for the third consecutive year, suggesting a steady pattern of problems. Despite massive education campaigns, antilock brakes, anti-skidding technology, rearview and side-mirror cameras, a phalanx of new exterior sensors, advances in vehicle-body design, and more airbags than ever, people are at higher risk of dying on US highways today than they were a few years ago. What changed?[36]

One factor might be the rising economy of the mid-to-late 2010s, which put more people on the road, making traffic heavier and accidents more likely. After all, the most commonly cited causes of traffic deaths haven't changed: speeding, driving without a seat belt, and drunk driving still top the list. But these factors can't explain the whole problem, because we are not only seeing an increase in the total number of deaths, but also in the number of deaths per mile driven, which controls for the number of cars on the road. Our roads aren't just busier; driving is becoming less safe.

Actually, it would be more accurate to say that *drivers* are becoming less safe. For years, virtually every new car available in the United States has been a safety and reliability marvel. Japanese automakers changed the game in the 1970s by bringing extremely reliable cars to market, forcing flagging US companies to catch up, which they eventually did. With just about every new car safer and more reliable than its predecessors, automakers began competing more on comfort and convenience. Earlier comfort features included things like plush seats, air conditioning, and power windows that rolled down at the touch of a button rather than the crank of a handle. These features tended to have limited impact on safety or improved it slightly by automating actions that could distract drivers.

More recent comfort and usability features, however, are focused on mobile media, communications, and information technology—technology designed to distract us. Some of these features are useful and may improve safety. GPS prevents us from getting lost and stuck in traffic, so that we are better able to avoid other cars and are less likely to get frustrated or to panic while driving. But new cars also brim with entertainment systems tethered to our smartphones. And those of us who don't have new cars still have smartphones. "It's not just talking on the phone that's a problem today," says Jonathan Adkins, executive director of the Governors Highway Safety Association. "You now have all these other apps that people can use on their phones."[37]

At this point, most people have some sense that distracted driving is a concern. Federal agencies trumpet National Distracted Driving Awareness Month. States have instituted fines for driving with a phone in hand. There are also calls to treat distracted driving as seriously as speeding, driving without a seat belt, and drunk driving.

Yet, despite all the attention, the problem is far worse than most of us realize. That's because the federal, state, and local governments largely rely on surveys, which vastly underrepresent the amount of distracted driving we engage in. Much more accurate data are available from Zendrive, a company that takes advantage of the technology in our smartphones—including the gyroscope, accelerometer, and GPS system—to monitor phone use alongside driving speed, location, and collision information. Not everyone has Zendrive installed, but the company has enough users to capture 100 billion miles of driver data. And the results of its analyses are concerning.

Drawing on data from 4.5 million users, Zendrive estimated in 2018 that close to 69 million drivers use their phone while behind the wheel every day in the United States. This figure is more than a hundred times great than what you will find in government statistics based on surveys. Zendrive's data also reveal that 60 percent of drivers check their phones at least once per trip. Most worrying, drivers who use their phone on the road average three and a half minutes of use for every hour of driving—driving fifty-five miles per hour, that is more than two miles or forty-two football fields while nearly totally distracted. The result is that smartphones are involved in a staggering number of crashes. Other research suggests that more than one in four drivers were on a phone just before a collision took place, and phone distraction occurs in more than half of car trips that result in a crash.[38]

Zendrive's 2019 "Distracted Driving Study" shocked again. They found that levels of distraction nearly doubled in one year. In addition to this increase, the report identified a group of particularly heavy users they labeled "phone addicts." These drivers "pick up their phones four times more than the average driver, use their phones six times longer than the general population, and are on the road longer than any other category of drivers." Zendrive's phone addicts are so distracted that the company suggests that they replace drunk drivers as the greatest threat to safety on the road.[39]

This assertion is backed by other research; it has been clear for some time that phone distraction can harm driving skill even more than drinking does. A 2014 study by the British Transportation Research Laboratory measured the reaction times of drivers when encountering an unexpected obstacle and found that drivers impaired by alcohol at the UK legal limit experience a 13 percent increase in reaction times compared to sober, nondistracted drivers. Drivers high on cannabis experienced a 21 percent increase in reaction times. And drivers texting or otherwise using smartphones? Reaction times were between 37 percent and 46 percent slower than those of an undistracted driver, making for an astounding threefold increase in reaction time compared with drunk drivers.[40]

Further confirmation of the perils of distracted driving comes from research using video cameras and car-mounted sensors. A 2016 study used these high-tech methods to explore the risk factors and causes associated with more than 900 accidents over a three-year period. The analysis includes 3,500 drivers and finds that they were engaged in some form of distracting activity 50 percent of the time. Distractions that took their eyes off the road imposed the highest risk. There are lots of distractions in the car, but handheld electronic devices stand out, resulting in a nearly fourfold increase in the risk of an accident. The authors conclude that texting and other mobile phone use while driving is the "single factor that has created the greatest increase in U.S. crashes in recent years."[41]

The response to the epidemic of distracted driving has largely been to ban the use of handheld devices in cars, but hands-free does not mean distraction-free. True, when using a hands-free device, we need not take our eyes off the road or hand off the wheel for as long as we otherwise would. But the attentional demands of talking on the phone while driving are quite high, even with a hands-free system.

Why? Research by the American Automobile Association suggests that common voice-controlled tasks and talking on the phone while driving are much more cognitively demanding than natural conversation with someone in the car. That's because a passenger, unlike a person on the other end of a phone call, is in the car, experiencing the same conditions as the driver. Passengers tend to slow or stop talking during critical driving moments and to point out upcoming turns and distractions. Passengers certainly can be distracting, but often they provide a second set of eyes that actually helps drivers be more responsive.[42]

When you think about it, it seems absurd that anyone would allow a phone to distract them while driving, yet people do it all the time. As one driver quoted in the 2019 Zendrive report explained, "I really do not think that using my phone is safe at all but for some reason I keep coming back to it." That is the power of habits formed in one environment and translated to another one. Phone distraction on the road reflects how deeply habituated we are to both driving and smartphone use. Unless you're new to the road, driving is second nature. As Neale Martin, a market researcher and former addictions counselor, puts it, "How can we be essentially behind the wheel of a guided missile going 60, 70, maybe 80 miles per hour—we are mere seconds and one distracted driver away from a potentially lethal accident—and what we feel is boredom?" Yet we are indeed bored because of the ordinary and habitual be-haviors associated with driving. And as we discussed, we often turn to our smartphones as a mood regulator when we are bored because the habit of checking them is powerfully ingrained.[43]

Layer on top of this the widespread but erroneous tendency to believe that we are good at multitasking because it makes us work harder in order to manage cognitive bottlenecks, discussed in chapter 5. In surveys Millennials admit to more "bad behaviors behind the wheel" than Baby Boomers do, including texting and driving. But Millennials are also twice as likely to report that they are "confident" in their ability to multitask with phones while driving com-pared with Baby Boomers (68 percent versus 34 percent). No data bear out this confidence.[44]

Not to let Boomers off the hook. They may not admit they text and drive, but the Zendrive data make clear that self-report surveys are highly unreliable. In some cases, survey-takers lie or underestimate their levels of risk-taking, but that is not necessarily the case here. The habit of multitasking with smartphones is so strong that we may not even realize we are doing it, including while driving. The habit will typically form while doing tasks that impose low cognitive demands, like watching television at home or talking to a friend. Then the habit becomes hijacked without our awareness as we shift to a more complex cognitive task, like navigating the roads. We associate our smartphones with the safer context in which we typically use them, and this association remains even when the shift in environment results in dramatically greater risks. As we saw earlier in the book, it is possible to be so distracted by a smartphone that one can injure oneself while walking or negotiating stairs. But use while driving is far more dangerous, and errors can be catastrophic.

Breaking this deadly habit won't be easy. As Martin reminds us, "We have this delusion that our conscious brain is in control and that if I think something, I will do it." But habits, and even more so addictions, in fact control us. This doesn't mean we are powerless, though. One avenue for change is collective action via lawmaking. We can and should lobby legislators to do more than pass relatively ineffective hands-free laws and to enforce existing and future rules. And because legal change comes slowly, we should apply social pressure and work to correct these dangerous habits in ourselves, our friends, and our loved ones.[45]

First, as with any habit or addiction, we need to recognize that we have a problem and commit to change. Second, pick a specific date to start breaking the habit and give yourself sufficient time to stay focused on the task. Change won't come immediately. There is a widespread misbelief that it takes three to four weeks to form a healthy habit or break an unhealthy one. In fact, the time needed is highly variable and usually considerably longer than this. Because the stakes are so high when it comes to distracted driving, you need to stick with your plan.[46]

Third, create a short pre-drive ritual, a kind of checklist to run through to ensure that you are prepared for your trip before you put the car in gear. This may include setting up your entertainment playlist or setting the radio to your favorite station. If you are driving to an unfamiliar location and need to rely on GPS, set your destination before driving and ensure that it will provide voice prompts. Voice prompts help us keep our eyes on the road instead of on our screens.

Fourth, let's put technology to good use here and make it part of the solution. Apple and Android phones can tell when we are driving and can automatically turn on do-not-disturb features that prevent notifications from reaching us. These features can be customized so that we can automatically reply to texts with a preferred message: "Sorry, I am driving right now and will respond to you shortly" is a good option. We can even set up these features to allow select contacts to push urgent messages through, but this setting should be used sparingly. There are also free apps that can help parents monitor and prevent distracted driving by their kids. Some of these systems disable certain phone features when the phone is moving above a set speed or alert parents when their teen drivers override do-not-disturb functions.

A final recommendation is to model good behavior as a driver and speak up as a passenger. Social pressure is a powerful force for change. If you are being driven by a distracted motorist, tell them their behavior makes you uncomfortable. As a passenger, be of help. Offer to navigate or control the entertainment choices. If the offender is your child or otherwise in your care, be firm: demand that they give you their phone while driving or do them the favor of locking it in the glove compartment. You can tactfully try the same approach with peers. We need to work together to keep in mind the long-term rewards of safer roads, reduced accidents, and passenger safety. Unlike other recommendations in this book, this one is a matter of life and death.

#10 Take a Real Break

The Greek mathematician, engineer, and inventor Archimedes was one of the great minds of Western antiquity. He anticipated modern calculus with his geometric theorems and derived one of the first accurate approximations of pi. However, Archimedes is perhaps best known for exclaiming, "Eureka!" after discovering his principle of buoyancy.

Passed down through the ages, the story features a local lord named Hiero who commissions Archimedes to help him with a problem. Hiero suspects that his new crown is not pure gold, as promised, and may in fact be plated silver. He wants to find out what the crown is made of but does not want anyone to cut into or harm the crown in the process of learning the truth about its composition. Archimedes takes on the challenge but is initially stumped. After several days of wrestling with the problem, he decides to take a break at a public bathhouse.

While relaxing in one of the tubs, he notices that his own body weight displaces the water in the tub. Because he knows that gold weighs more than silver, he figures he can use a precise measurement of water displacement to calculate the mass of the crown and compare what he finds to the known masses of gold and silver and thereby learn the true nature of Hiero's crown. This was the famous eureka moment (albeit one that historians debate). The point of Archimedes's tale stands the test of time, though: inspiration often strikes us when we are in a relaxed state of mind.[47]

Modern psychologists and neuroscientists are only beginning to understand why this is the case. Benjamin Baird is one of these psychologists. A research scientist in the School of Medicine and Public Health at the University of Wisconsin, he studies approaches to problem-solving and associations between mental states and creativity. Inspired by the counterintuitive findings of prior research, he sought to find out whether mind wandering not only enhances creativity but may even be more useful for problem solving than is focused deliberation.

Baird's study looked at 145 adults, aged 19–32, who were randomly assigned to work on a set of challenges known as the Unusual Uses Task, a widely used measure of divergent thinking and creativity. The task requires participants to generate as many unusual and unique uses as possible for a common object, such as a brick, in a limited amount of time. The key to Baird's study is the break period before carrying out the task. Some participants were assigned to do something easy before taking on the Unusual Uses Task, which encourages mind wandering, while others were assigned to do something cognitively demanding, which minimizes mind wandering.

The study found that taking a break with an easy task that facilitates mind wandering results in significantly better scores in problem solving and creativity. This research does not explain why mind wandering enhances creativity and problem solving, but Baird and his colleagues do suggest a brain mechanism that may be responsible: the default mode network.[48]

The default mode network refers to a group of interconnected brain structures that work on both a conscious and nonconscious level when we are awake but resting. Unlike most brain networks, this network is increasingly activated when external task demands are low. The default mode network was discovered by accident as neuroscientists began to analyze data from diverse brain-imaging studies. Such studies often ask participants to lie quietly and "just relax" or "let your mind wander" before they begin doing whatever task is required for the purposes of the research. At first these instructions were used to simply calibrate the neuroimaging machine prior to data collection.

But over time neuroscientists realized that a consistent and remarkably robust brain network was activated during this period when participants were awake and at rest.[49]

Neuroscientists now believe this network underlies the internal monologue we experience when we are engaged with nondemanding cognitive tasks or are in a relaxed state of mind. Tasks that demand more brain resources in the form of attention, emotion, or memory shift the brain away from this network accordingly. But when we resume a state of resting wakefulness with no clearly defined task, this unique network is consistently activated.[50]

Interestingly, the prefrontal cortex plays a prominent role in this network. Just prior to a moment of insight, areas of the prefrontal cortex involved in mind wandering become activated in preparation for the birth of an idea. Returning to our metaphor of the prefrontal cortex as the conductor in the brain's symphony, imagine our default brain mode as the warm-up period before the orchestra begins to play music. The musicians spend a few moments on stage gathering themselves. Some are lightly active, reviewing their score, setting the pages in order, tuning their instruments. Others chat or simply sit quietly. But when the conductor calls the players in the orchestra to attention, every one of them is silent and prepared to produce beautiful music. The brief period of relaxed preparation before the conductor taps their baton is essential to the performance. Within this resting state, the mind can wander until the prefrontal cortex calls for attention.

While more research is needed, there is evidence to suggest that the more we allow our minds to wander, the more creative our insights are and the more flexible our thinking becomes. This speaks to the importance of taking breaks, particularly in our hyperconnected, 24/7 world, to let our minds wander in a state of relaxation. But taking a break doesn't mean we should set aside work to check email, social media, and news alerts on our phones. These tasks are often cognitively less demanding than studying for an exam or handling a work task, but they still keep us focused. Taking a real break means allowing the mind to wander.[51]

Unfortunately, achieving this truly disengaged state seems to be quite hard for many of us, and it may well be that the myriad temptations of the digital age are contributing to this difficulty. In 2014, researchers at the University of Virginia undertook a series of experiments to gauge whether people could adopt a relaxed state of mind and whether the experience was aversive or appealing. The team first recruited young adults to spend a short period of time—six to fifteen minutes—in an unadorned room with the instruction to entertain themselves with their own thoughts. Participants were required to stow any belongings, including smartphones and other media technology, outside and were asked to stay awake and alone in their seats. Nearly 90 percent achieved some mind wandering, but almost half reported that they did not enjoy the experience.

Concerned that the unfamiliar environment was a factor in participants' lack of enjoyment, the team carried out another experiment in which young-adult participants did the same exercise but this time at home. In this case, nearly one-third of participants reported that they cheated by getting out of their chair or by using their smartphone. In another variant of the study, participants were allowed to read a book or listen to music but could not communicate with anyone. Here, perceived enjoyment went up slightly. Similar results were found in a version of the study carried out with older adults.

In a final variant of the study, researchers tweaked the rules somewhat and gave the participants a new, disconcerting option. As in some of the other experiments, participants were instructed to sit quietly in the unfamiliar lab without any belongings or tasks. But this time, all were required to sit for the full fifteen minutes unless they ended the experiment by self-administering a mild but unpleasant electric shock. Astonishingly, two-thirds of the men and one-quarter of the women in the study chose the shock over sitting quietly with their own thoughts.[52]

This was not a study of modern media, communications, and information technology, although we can draw some relevant inferences from the findings. In particular, reading books and listening to music only slightly increased people's enjoyment of sitting alone, suggesting that these activities are perceived as little better than doing nothing. Meanwhile, cheaters often turned to their smartphones, not to books. But the central insight is clear enough: people prefer doing something, anything—including inflicting pain on themselves—rather than letting their minds wander while doing nothing. Almost any stimulus is preferable to no stimulus at all.

Fortunately, there are lots of ways to facilitating mind wandering, if we will allow ourselves to try. Some we can't really avoid: light housework such as folding clothes or washing dishes can be an opportunity to disengage. We've all heard of people who say they do their best thinking in the shower; the simple and habitual task of cleansing our bodies in water facilitates mind wandering for many people. More deliberately, one might go for a walk, preferably in nature.

For our purposes, let us focus on two disengagement activities that have been shown to contribute to brain health: meditation and physical exercise. We have seen how a broken tech-life balance weakens the prefrontal cortex and hijacks our reward centers. These two activities can help us repair the damage. Clear scientific evidence shows that meditation enhances functioning in the prefrontal cortex, while exercise can help reorient our reward centers toward healthy behaviors.

Most variants of mindful meditation were adopted from Buddhist traditions that, until the last few decades, were practiced primarily in India and in specialized retreat centers scattered around the world. But these Buddhist-inspired meditation practices have been present in Western cultures for many years and have lately grown dramatically in popularity. Many experts believe that meditation is taking off because it is an antidote to the ills of an age in which technology drives us to distraction. People seeking out meditation are looking for something they cannot find in ordinary life: stillness. And with stillness, healing and enhanced brain function. Indeed, with considered practice, the benefits of meditation are legion. Even modest amounts of meditation can lead to significant reductions in stress and chronic pain as well as reduced depression, anxiety, and fewer negative ruminations, among other desirable outcomes.[53]

Seeking to understand the power of meditation, neuroscientists have been investigating its effects for over two decades. Numerous compelling findings suggest that meditation changes the brain in ways that can offset some of the rewiring brought on by our mobile media, communications, and information-technology habits. Not surprisingly the prefrontal cortex, one of the main brain regions associated with focused attention, is involved in meditation. Research has shown that even a short-term meditation practice can increase blood flow to the prefrontal cortex and other areas of the brain related to emotion regulation. Indeed, assorted meditation practices affect a range of brain networks, suggesting that certain meditation approaches might be well suited to relieving stress and others anxiety or depression.[54]

Meditation is another arena in which the very technology that gives rise to many problems offers some solutions. A growing number of meditation apps on mobile devices deliver a plethora of choices for beginners and experts alike. These apps provide soothing background music and nature sounds to help with relaxation and timed guidance in assorted calming voices to help guide practitioners. My recommendation is to try several of them and experiment until you find the right approach that suits your temperament and style.

As a perfect complement to meditation, physical exercise provides another real break that offers many advantages while allowing opportunities for mind wandering. The health benefits of physical activity are well documented and almost too numerous to list. Regular exercise lowers our risk of death from heart disease, diabetes, and stroke and has been shown to reduce the incidence of certain cancers. As we age, regular exercise also helps maintain healthy bones and muscle mass, expand lung capacity, reduce the risk of arthritis and falls, and keep our weight in check.

None of this is news. What is less well known and has been discovered more recently is that regular exercise has significant benefits for our brains and mental health as well. Research shows exercise can increase resiliency, improve our moods, lower our anxiety, and counteract depression. Exercise also boosts brain power in all of the areas harmed by media multitasking, enhancing our ability to sustain attention and increase our executive functioning. Even short exercise routines, as brief as ten minutes a day, have been proven to boost attention and concentration.[55]

While some of the benefits of exercise take a while to set in, others are more immediate. These immediate benefits are natural rewards in the form of a dopamine hit. An effective workout produces elevated levels of dopamine and other neurochemicals, contributing to positive post-exercise feelings. The increase in dopamine is so significant that exercise has been shown to reduce cravings for several drugs of abuse. For good reason, exercise is even being prescribed as an adjunct in addiction treatments.[56]

Both exercise and meditation provide the break we need to let our thoughts wander and help us focus anew. These may sound like goals at cross-purposes, but in truth they are not. We need to let ourselves go sometimes in order to refresh and return to our daily lives with renewed vigor and attention. We also need the space to be mindful, so that we can implement these recommendations and retake the control that we have lost to our technology habits. Beneficial habits, like considered meditation practice and regular exercise, help us orchestrate our lives. In the process, we will make a few wrong turns and we will make mistakes. But there is no failure when we work hard to institute positive habits; there are only opportunities to learn. The more we focus on the impact of our use of technology day to day, moment by moment, the more mindful we become. The more mindful we become, the better our brains will work on behalf of ourselves and our loved ones.

In the end, we may not be as smart as Archimedes, but even the ancient Greeks knew that taking a break is an excellent way to facilitate eureka moments—at least according to legend. And while there are plenty of recommendations out there on how to take short breaks at work—go outside for a breath of fresh air, take your eyes off your screen for a few minutes between tasks—retraining our brains requires more time and effort than that. Many of life's natural rewards don't come cheaply or easily. Achieving them requires planning, especially given the obstacles of our contemporary digital lives. But making the effort is increasingly important, as our technology habits rewire our brains against our best interests. Fortunately we can take care of our brains even amid the modern maelstrom by turning to meditation and exercise—two of humanity's oldest pursuits, now backed by rigorous scientific study.

10

IS THERE HOPE?

There are so many causes for worry about the corrosive effects of technology in our lives and on our brains. But there are also reasons to believe that we can use digital literacy to alter our relationship with mobile media, communications, and information technologies. We know that humans are capable of positive change because we have seen it before. I conclude with four clear signs of hope.

We Are Getting Smarter

The late James Flynn spent much of his career studying the trajectory of human intelligence. His extensive research suggests that human cognition and mental habits have changed dramatically over the last century, in ways that result in impressive increases in IQ test scores. This increase in measured intelligence has been seen across thirty-four countries including the United States, Australia, Japan, South Korea, and most of Europe. It appears that we are getting smarter as a species.[1]

Flynn argued that rising test scores reflect the increasing intelligence required to handle the growing complexity of our world in the twentieth century. In the past, the pace of life and work was slower, education was harder to access, and technology was less sophisticated. Modern life, however, placed new demands on cognition and forced changes in the ways we think about the world around us. We had to classify more diverse objects, master more problem-solving skills, assume the burden of increased hypothetical and abstract thinking, and engage in more scientific reasoning. We have adapted to these demands with more high-quality educational and training opportunities. Notwithstanding the seemingly endless debates about how best to teach children, over the past hundred years a higher percentage of people across the globe have had the opportunity to finish high school, go to college, and obtain a graduate degree.[2]

Flynn's analysis of IQ scores was groundbreaking. It was also stunningly simple. He realized that increases in measured intelligence were being masked by a methodological quirk. Every few years since they were originally developed, IQ tests have been revised so that the median score is always 100 points. Flynn found that, looking back to the early tests, every adjustment resulted in an increase in the raw score needed to achieve the median score—test-takers had to keep doing better and better in order to be recorded as having middle-of-the-road intelligence. The adjustments averaged around three points per decade, a shift that added up over time. "We don't just get a few more questions right on IQ tests," Flynn said. "We get far more questions on IQ tests correct than each successive generation back to the time when they were invented."

If we take the scores of people from the early exams decades ago and normalize them to the modern test, their average IQ score would be seventy—the modern-day cutoff for mild intellectual disability. Similarly, if we take the scores from today's tests and normalize them to tests from a hundred years ago, our average score would be 130, the cutoff for a gifted intellect. This means that if a group of people today took the same IQ test from a hundred years ago, the average person would qualify for Mensa membership. This massive increase in measured intelligence over the course of generations is now known as the Flynn effect.

Experts vigorously debate the possible explanations for this increase in IQ. Some speculate that improvements in health, nutrition, and early-childhood education are factors. Others point to changes in the character of education, in particular the shift in emphasis away from rote memorization and toward problem-solving skills. Another factor may be increased access to scientific reasoning and knowledge. The groups being measured are also dynamic; for instance, changing immigration patterns may have affected the overall composition of test-takers in ways that influence IQ at the population level. Our jobs have also become more cognitively demanding, simultaneously influencing choices surrounding education and giving us opportunities to exercise our intellectual capacities. In 1900 only 3 percent of American workers had cognitively demanding jobs, working, for instance, as teachers, lawyers, or doctors. Today 35 percent of Americans work in fields that are cognitively demanding, fields that did not exist in 1900 such as allied health care, consulting, and computer science.[3]

Recent analysis suggests the Flynn effect may have stalled over the last couple of decades. It is possible we have reached our capacity as a species. Or maybe new habits around media, communications, and information technology are having an impact. Still, we can take hope from the realization that society as a whole can make cognitive gains, and these gains are almost certainly a product of the environment we create for ourselves. The smarter we are on average, the better our chances of navigating the many challenges that face us.[4]

Young People Are a Source of Hope

The cognitive and emotional deficits associated with rising media multitasking and social media use are a source of serious worry. But we should also recognize some of the good that mobile media technology enables. One of the great benefits is that children growing up today have access to more information and more points of view than ever before. This access presents challenges, as I have described in detail. Young people recognize these challenges, too: surveys of US teens by the Pew Research Center find widespread concerns about social pressure and excessive online "drama." However, a large majority of teens also credit social media with helping them build stronger friendships and exposing them to a more diverse world.[5]

Mobile media, communications, and information technologies are, after all, tools. We can use them for good and for ill, developing connections and understanding even as we may sacrifice attention and emotional well-being. Likewise, the elegant neurobiology of the immature teen brain, which simultaneously increases specialization in functional areas and becomes selectively more interconnected, is a two-sided coin. The immature prefrontal cortex of teens results in a lack of self-control that puts them at considerable peril. Yet the same disinhibition leaves teens more open to creativity and potentially beneficial risk-taking.

Adolescence is our time to experiment, on behalf of ourselves and our communities. It is a period in which each generation learns to deal with novel environmental factors, such as major technological and other changes that adults often have a harder time managing. As developmental expert Jay Giedd puts it, "The link between adolescent brain evolution and the digital revolution does not lie in a selection pressure wherein those with greater capacity to handle the demands of the technological changes have greater reproductive success. The link lies in the evolutionary history that has made the human adolescent brain so adaptable."[6]

In other words, it is young people, today's digital natives, who are adapting to the digital revolution on behalf of us all. Their inclination to experiment and creatively approach risk makes them valuable guides through this new technological terrain. Adolescents today are using mobile devices to explore new frontiers in new ways. They are more open to diversity than are older adults. One could make the case that a willingness to share stories and ideas across the borders and boundaries of politics, language, geography, and culture will, over time, foster more tolerance and understanding of unfamiliar people and could contribute to a reduction in racism and dehumanization so pervasive around the world. The same adolescent brain that struggles mightily controlling impulses and coping with emotional hardship is also uniquely driven toward exploration, and that is something we will need in order to meet our collective challenges and embrace the world around us on more humane terms.

For most of our time on this planet, the average life expectancy was about thirty years, much of it spent in adolescence. While this period of neuronal development comes with risks, adolescence is also a special period of brain growth that confers tremendous capacity for adaptability, imagination, and creative problem solving. Humans and their early relatives have used this developmental stage to respond to extraordinary environmental challenges for over a million years. This offers us hope for the future.

The Shrinking Digital Divide

We wouldn't be talking about the challenges of living with the digital revolution if we hadn't solved one of the earlier vexing problems associated with modern media, communications, and information technology: getting it into people's hands. At the turn of the millennium, nearly half of all Americans were offline. Political and civic leaders raised legitimate concerns that lower-income and less-educated people, particularly minorities, would be left out of the emerging digital economy. Less than two decades later, in 2018, nearly 90 percent of all Americans had access to the internet, and 95 percent of US teens had access to a smartphone.[7] The same trend has played out globally. About 94 percent of the world's population is now able to receive a mobile phone signal if they can afford it, and the total number of mobile phone users is estimated at between 5 and 7 billion people.[8]

Some disparities in usership linger. Nearly one-third of Americans aged sixty-five and older do not access the internet, and income differences continue to be notable. Americans making $30,000 or less have lower rates of internet use and less access to broadband networks. At the same time, race and gender gaps in use and access have closed significantly. As one Pew report puts it, "Groups that have traditionally been on the other side of the digital divide in basic internet access are using wireless connections to go online."[9]

This is another source of both hope and challenge. More smartphone usership means more opportunities to fall into the traps set for us by clickbait, compulsion loops, social comparisons, online narcissists, and misinformation. This leads to more distraction, divisiveness, and depression as we compromise productivity and meaningful social interactions. Yet as we cross the digital divide, the ubiquitous access to information and enhanced educational opportunities balance the ledger to some extent. It has never been my claim that our modern technology is inherently bad for us, only that we often use it unwisely because we have focused too much on titillation and too little on the consequences of our lack of digital literacy. In spite of this, the promise of this technology is real. It can help make life better for all of us.

Technology Can Be Part of the Solution

From Shenzhen to Singapore, Tokyo to Tel Aviv, Boston to Austin, and many places in between, academics, entrepreneurs, and technology companies fueled by venture capital are racing to develop new software and hardware that can help guide us through the next chapter of the digital age. This is critical. If technology is going to be part of the solution, then technology creators need to start thinking differently, focusing less on hijacking our brains and more on empowering users.

We are beginning to see a shift in this direction—toward technology that enables better living rather than unhealthy habits. For example, there are startups tackling the need for parental controls, to help manage kids' screen time and other technology use. Companies are making apps that use rewards to motivate people to lead healthier and more active lives. Of course, we need to maintain some perspective when it comes to health and fitness apps. The quality of disease-management apps has improved, but many makers of health and fitness apps—there are more than 350,000 on the market as of this writing—advertise bold claims that far outstrip any evidence of effectiveness.[10]

Another positive development comes from the very companies that got us hooked on our smartphones in the first place. Major firms, including Apple and Google, are doing more to help us manage screen time and achieve tech-life balance. They are introducing features that limit time spent on devices and apps, reduce distractions from notifications, and enable more parental control over children's usage.[11] This suggests that public criticism and concerns about technology use may be having an effect. Companies also fear government regulation and may be proactive for that reason. Whatever is motivating tech giants to promote "digital wellness," it is in the best interest of users and therefore probably of the companies themselves.

Many of the benefits of corporate responsibility—and the public pressures that nurture it—may lie in the future. There is much more technology development to come, which will change the capabilities of the mobile powerhouse in our pockets we euphemistically call the smartphone. As artificial intelligence and augmented reality technologies improve, they too will become part of our lives, creating new opportunities but also inviting new dilemmas. Changing public attitudes and promoting digital literacy today can set us on a better course tomorrow. Ideally, every new piece of hardware and software will be launched with healthy habits in mind. We have an opportunity to influence legislators, regulators, and the marketplace to be more proactive than reactive, to help prevent some of the problems that otherwise lie ahead.

The Human Brain Is Capable of Change

Finally, each of us is endowed with what is still the most advanced technology on the planet—a highly adaptable organ that can change to meet shifting demands across the lifespan. Our brains are not static after the explosive growth of early childhood and the unique period of adolescence. Even adult and elderly brains are constantly making new neurons and neuronal connections in response to learning, genetic instructions, and changes in the environment.

Neuroplasticity, the formation of new neuronal connections in the brain, is a never-ending process. Neurons are constantly sending out tentacle-like structures called dendrites that bridge the gap with other neurons. The creation of these new neuronal connections, known as synapses, is the foundation of learning. At the same time, our brains edit themselves through apoptosis, the diminution and eventual death of cells, including neurons. Throughout our lives, neurons die off in a coordinated fashion. These constant processes of formation and elimination influence our feelings, thoughts, and behaviors, allowing us to respond to the stimuli and ceaseless changes in the environment around us.

In addition to new synapses, the brain can also create new neurons. Scientists used to believe that the adult brain reached a fixed number of neurons that inevitably decreased with age. However, new neuroscience clearly suggests that the brain not only is capable of neurogenesis but that it happens as a matter of course. The hippocampus, a specialized area of the subcortex critical for certain types of memory and learning, makes thousands of new neurons a day, fostering the potential for lifelong learning. But there is a catch: most of these neurons die within a few weeks of their birth.[12]

Why? We know from animal studies that neurogenesis follows a simple rule: "use it or lose it." In this context, use means effortful learning, requiring multiple trials, focused attention, and time spent acquiring knowledge or skills. Skimming materials with divided attention or simply going through the motions at work or in school is not sufficient. The payoff of effortful learning includes not just the new habits and skills one learns but also the benefit of new neurons in the brain that increase resiliency against neurocognitive decline—the loss of memory and brain function as we age. The more we use our brains in healthy ways, the more our neurons respond in healthy ways. But the reverse is also true. Neuroplasticity is a double-edged sword—a key to realizing our potential and also the reason we are all susceptible to unhealthy habits and addictions over time.

Our incredibly adaptive brains will see us through the technology revolution that we are currently experiencing, a revolution that is likely to accelerate over the coming decades. But there is a difference between surviving and thriving. We will need to be proactive rather than reactive if we are to take advantage of this accelerating revolution and all that it has to offer.

This is why I talk about the need for digital literacy and tech-life balance. There is so much benefit to be derived from the fruits of the digital age, and there will be great possibilities inherent in whatever technologies come next. But we have a lot to lose as well, unless we can make good use of our tools without succumbing to their control. We won't do that if we remain stuck in a highly reactive mode. Twenty years after the mass adoption of the internet and fifteen years after the emergence of the smartphone, we are still seduced by the superficial. We respond to every app ping and message ring reflexively, mindlessly changing our behaviors on a massive scale, with very real consequences for our brains and our children's brains. We suffer the tyranny of titillation and are hooked by superstimuli, resulting in habits that do us harm. At the same time, we are prey to endless distractions that diminish our productivity, keep us from important relationships, undermine our mental health, and in certain contexts, put us in serious physical danger.

Some might reply that we are reward-seeking machines, evolved to experience what pleasures we can. If compulsion loops make us feel good, why not embrace them? Yet, while we are built for rewards, we are equally built for balance. Our neurological reward system is complemented and constrained by our prefrontal cortex, the conductor in our own personal symphonies. As we have seen, the prefrontal cortex is a product of our evolutionary history and our environment, whose purpose is in part to prevent us hurting ourselves and our communities by halting us from going all in on momentary pleasures. Human beings are not just built to give in to whatever delights come our way. We are also built to balance these rewards against the more fulfilling but less immediate benefits of learning, wisdom, compassion, and companionship.

Now it is time to be proactive and protect our precious brains. With planning, foresight, and the guiding light of thoughtful scientific investigation, we can collaborate to put humanity at the center of the technology revolution. No one segment of society is going to get the job done on its own. We need leaders in the technology industry to recognize the importance of user control and human-centered design. We need educators to teach digital literacy. We need governments to support curricula, messaging, and appropriate regulation of industries whose profits too often come at great risk to the public.

And all of us can individually commit to some simple steps toward orchestrating our own tech-life balance and protecting young people in our lives from the most serious effects of new mobile media, communication, and information technologies. As we evaluate the role of smartphones, social media, and other digital wonders in our own lives, we must humbly accept that technology itself has no agenda. All technology is like the proverbial knife. In the hands of a murderer, a knife is a weapon of destruction and death; the same tool in the hands of a surgeon brings healing and life. The difference lies in intent and motivation. Technology should be a tool we use, not a tool that uses us. It is that simple.

NOTES

ACKNOWLEDGMENTS

INDEX

NOTES

Introduction

1. Abida Sultana et al., "Digital Screen Time during COVID-19 Pandemic: A Public Health Concern," unpublished manuscript, *SocArXiv* (online only), September 1, 2020, https://doi.org/10.31235/osf.io/e8sg7.

2. Nielsen Company, "COVID-19: Tracking the Impact on Media Consumption," Nielsen Insights, June 16, 2020, https://www.nielsen.com/us/en/insights/article /2020/covid-19-tracking-the-impact-on-media-consumption; Nielsen Company, "How Streaming Enablement in 2020 Has Changed the Media Landscape," Nielsen Insights, October 6, 2020, https://www.nielsen.com/us/en/insights/article /2020/how-streaming-enablement-in-2020-has-changed-the-media-landscape; Matt Richtel, "Children's Screen Time Has Soared during the Pandemic, Alarming Parents and Researchers," *New York Times,* January 17, 2021.

3. Alison Abbott, "COVID's Mental Health Toll: How Scientists Are Tracking a Surge in Depression," *Nature News,* February 3, 2021, https://www.nature.com /articles/d41586-021-00175-z.

1. Media Matters

1. Samantha Tatro and Omari Fleming, "Man, Distracted by Electronic Device, Identified after Falling to Death at Sunset Cliffs," NBC News, San Diego, December 25, 2015, http://www.nbcsandiego.com/news/local/Man-Dies-After-Falling -Off-Cliff-at-Sunset-Cliffs-Lifeguards-363534491.html; "Japanese Tourist at Taj Mahal Dies after Fall," *BBC News India Online,* September 18, 2015, http://www.bbc .com/news/world-asia-india-34287655; Rita Sobot and Ruth Halkon, "Man Taking a Selfie Is Gored to Death by Bull in Front of Horrified Crowd," *Mirror,* August 10, 2015, http://www.mirror.co.uk/news/world-news/man-taking-selfie -gored-death-6225812; Suzan Clarke, "Texting Woman Falls off Pier into Lake

Michigan," ABC News, *Good Morning America,* March 23, 2012, http://abcnews
.go.com/blogs/headlines/2012/03/texting-michigan-woman-falls-off-pier-into
-lake; "Girl Falls in Mall Fountain while Texting," January 18, 2011, https://www
.youtube.com/watch?v=umRXAkZ8Xoo; "Texting Lady Explains How She Fell
in Mall Water Fountain," January 22, 2011, https://www.youtube.com/watch?v
=pcRxWUXKC88.

2. Ford Motor Company, "Looking Further with Ford," 2017 Trend Report, n.d.,
https://media.ford.com/content/dam/fordmedia/North%20America/US/2016
/12/7/2017-Looking%20-Further-with-Ford-Trend-Report.pdf.

3. Todd Gitlin, *Media Unlimited: How the Torrent of Images and Sounds Overwhelm
Our Lives* (New York: Metropolitan, 2001; New York: Holt Paperbacks, 2007),
72–77.

4. The Nielsen Company, "2015 Total Audience Report Q4," accessed October 1,
2019, http://www.nielsen.com/us/en/insights/reports/2016/the-total-audience-
report-q4-2015.html; Jason Lynch, "U.S. Adults Consume an Entire Hour More
of Media Per Day Than They Did Just Last Year: For a Daily Total of 10 Hours, 39
Minutes," AdWeek, June 27, 2016, http://www.adweek.com/tv-video/us-adults-
consume-entire-hour-more-media-day-they-did-just-last-year-172218/; The Nielsen
Company, "2018 Total Audience Report Q2," accessed October 1, 2019, https://
www.nielsen.com/us/en/insights/reports/2018/q2-2018-total-audience-report.
html; Jack Loechner, "TV Screen Dominates Adult Viewing," Media Post Research
Brief, June 12, 2017, https://www.mediapost.com/publications/article/302658
/tv-screen-dominates-adult-viewing.html.

5. Jack Loechner, "TV Screen Dominates Adult Viewing," research brief, *MediaPost*
(online resource for advertising professionals), June 12, 2017, https://www
.mediapost.com/publications/article/302658/tv-screen-dominates-adult
-viewing.html.

6. Jack Wakshlag, email to author, June 5, 2017.

7. Nick Bilton, "Apple Is the Most Valuable Company," *New York Times,* August 9,
2011; Thomas Heath, "Apple Is the First $1 Trillion Company in History," *Wash-
ington Post,* August 2, 2018; "U.S. Smartphone Penetration Surpassed 80 Percent
in 2016," blog post, Comscore (media analytics company), February 3, 2017,
https://www.comscore.com/Insights/Blog/US-Smartphone-Penetration
-Surpassed-80-Percent-in-2016.

8. Michael DeGusta, "Are Smart Phones Spreading Faster than Any Technology
in Human History?" *MIT Technology Review,* May 9, 2012, https://www.technology
review.com/s/427787/are-smart-phones-spreading-faster-than-any-technology
-in-human-history.

9. George P. Slefo, "Desktop and Mobile Ad Revenue Surpasses TV for the First
Time," *AdAge,* April 26, 2017, http://adage.com/article/digital/digital-ad-revenue
-surpasses-tv-desktop-iab/308808.

10. Steve Hasker, "The World According to Nielsen," April 30, 2014, https://www
.youtube.com/watch?v=9O3vETqRW8I.

11. Deloitte Center for Technology, Media & Telecommunications, "How the Pandemic Has Stress-Tested the Crowded Digital Home," 2021, https://www2.deloitte
.com/content/dam/insights/articles/6978_TMT-Connectivity-and-mobile
-trends/DI_TMT-Connectivity-and-mobile-trends.pdf.

12. Jack Loechner, "Media Consumption Grows, Enhanced by Internet," *MediaPost*
(online resource for advertising professionals), June 11, 2015, https://www.media
post.com/publications/article/251441/media-consumption-grows-enhanced-by
-internet.html.

13. Elihu Katz, Jay G. Blumler, and Michael Gurevitch, "Uses and Gratifications Research," *Public Opinion Quarterly* 37, no. 4 (1973): 509–523.

14. "Time Inc. Study Reveals That 'Digital Natives' Switch between Media Devices
and Platforms Every Two Minutes, Use Media to Regulate Their Mood," press
release, Business Wire, April 9, 2012, https://www.businesswire.com/news/home
/20120409005536/en/Time-Study-Reveals-%E2%80%9CDigital-Natives%E2
%80%9D-Switch-Devices. Participants signed a consent form and were paid to
participate. The equipment and cameras were turned off while participants were
at work, to avoid any sharing of confidential business information. Analyses focused on nonwork-related media consumption. Every second of the physiology
data was analyzed, and the camera data were hand-coded by trained raters for
every interaction with media devices. The media devices were divided into two
types: 1) nondigital or "traditional" media devices included television, radio,
newspapers, and magazines; 2) digital media devices connected to the internet
included personal computers, laptops, tablets, and smartphones. Media attention spans were calculated with reference to the average time one media platform
held a participant's gaze before switching to another.

15. Brian Steinberg, "Study: Young Consumers Switch Media 27 Times an Hour,"
AdAge, April 9, 2012, http://adage.com/article/news/study-young-consumers
-switch-media-27-times-hour/234008.

16. The second study was also conducted in the Boston metro area. It was funded
by Nielsen on behalf of the Council of Research Excellence, a consortium of
Nielsen clients.

17. Mariam Arain et al., "Maturation of the Adolescent Brain," *Neuropsychiatric Disease and Treatment* 9 (2013): 449–461, https://dx.doi.org/10.2147%2FNDT.S39776.

18. Marie E. Schmidt et al., "The Effects of Background Television on the Toy Play
Behavior of Very Young Children," *Child Development* 79, no. 4 (2008): 1137–1151,
https://doi.org/10.1111/j.1467-8624.2008.01180.x.

19. Betsy Frank et al., "A (Biometric) Day in the Life: A Cross Generational Comparison of Media Platforms," White Paper, Time Inc., May 2013, https://www
.innerscoperesearch.com/news_old/time_warner-whitepaper-2013.pdf.

20. "The New Normal: Parents, Teens, and Devices around the World," Common Sense Media, October 1, 2019, https://www.commonsensemedia.org/research /The-New-Normal-Parents-Teens-and-Devices-Around-the-World.

2. The Power of the Prefrontal Cortex

1. Eiluned Pearce, Chris Stringer, and R. I. M. Dunbar, "New Insights into Differences in Brain Organization between Neanderthals and Anatomically Modern Humans," *Proceedings of the Royal Society B* 280, no. 1758 (2013): 1–7, https://doi .org/10.1098/rspb.2013.0168.

2. Chet C. Sherwood, "Are We Wired Differently?" *Scientific American* 319, no. 3 (2018): 60–63, http://doi.org/10.1038/scientificamerican0918-60.

3. Robert T. Knight and Donald T. Stuss, "The Prefrontal Cortex: The Present and the Future," in *Principles of Frontal Lobe Function,* ed. Donald T. Stuss and Robert T. Knight (Oxford: Oxford University Press, 2002), 574.

4. M. Marsel Mesulam, "Behavioral Neuroanatomy: Large-Scale Networks, Association Cortex, Frontal Syndromes, the Limbic System, and Hemispheric Specializations," in *Principles of Behavioral and Cognitive Neurology,* 2nd ed., ed. M. Marsel Mesulam (Oxford: Oxford University Press, 2000), 47–48.

5. Joaquin M. Fuster, "Introduction," in *The Prefrontal Cortex,* 5th ed., ed. Joaquin M. Fuster (London: Academic Press, 2015), 8.

6. Mesulam, "Behavioral Neuroanatomy."

7. Michael Harre, "Social Network Size Linked to Brain Size: How and Why the Volume of the Orbital Prefrontal Cortex Is Related to the Size of Social Networks," *Scientific American Mind,* August 7, 2012, https://www.scientificamerican.com /article/social-network-size-linked-brain-size.

8. Matthias Meyer et al., "A High-Coverage Genome Sequence from Archaic Denisovan Individual," *Science* 338, no. 6104 (October 12, 2012): 222–226, https://doi .org/10.1126/science.1224344.

9. Pallab Ghosh, "Neanderthals' Large Eyes 'Caused Their Demise,'" *BBC News,* March 13, 2013, http://www.bbc.com/news/science-environment-21759233.

10. Sergey Gavrilets and Aaron Vose, "The Dynamics of Machiavellian Intelligence," *Proceedings of the National Academy of Science of the United States of America* 103, no. 45 (2006): 16823–16828, https://doi.org/ 10.1073/pnas.0601 428103.

11. Wendell Steavenson, "Ceausescu's Children," *Guardian,* December 10, 2014.

12. Harvard Center on the Developing Child, "Five Numbers to Remember about Early Childhood Development," accessed October 1, 2019, https://developingchild .harvard.edu/resources/five-numbers-to-remember-about-early-childhood -development/#cps.

13. Eric Courchesne et al., "Normal Brain Development and Aging: Quantitative Analysis at In Vivo MR Imaging in Healthy Volunteers," *Radiology* 216, no. 3 (2000): 672–682, https://doi.org/10.1148/radiology.216.3.r00au37672.

14. Christopher Bergland, "How Does Neuroplasticity and Neurogenesis Rewire Your Brain?" *Psychology Today* blog, February 6, 2017, https://www.psychology today.com/blog/the-athletes-way/201702/how-do-neuroplasticity-and -neurogenesis-rewire-your-brain.

15. Daniel J. Siegel, *The Developing Mind: Toward a Neurobiology of Interpersonal Experience* (New York: Guilford Press, 1999).

16. Robert J. Waldinger and Marc S. Schulz, "The Long Reach of Nurturing Family Environments: Links with Midlife Emotion-Regulatory Styles and Late-Life Security in Intimate Relationships," *Psychological Science* 27, no. 11 (2016): 1443–1450, https://doi.org/10.1177/0956797616661556.

17. Everett Waters et al., "Attachment Security in Infancy and Early Adulthood: A Twenty-Year Longitudinal Study," *Child Development* 71, no. 3 (2000): 684–689, https://doi.org/ 10.1111/1467-8624.00176.

18. Waldinger and Schulz, "The Long Reach of Nurturing Family Environments."

19. Charles H. Zeanah et al., "Institutional Rearing and Psychiatric Disorders in Romanian Preschool Children," *American Journal of Psychiatry* 166, no. 7 (2009): 777–785, https://doi.org/10.1176/appi.ajp.2009.08091438; Charles H. Zeanah et al., "Attachment in Institutionalized and Community Children in Romania," *Child Development* 76, no. 5 (2005): 1015–1028, https://doi.org/10.1111/j.1467-8624 .2005.00894.x.

20. Margaret A. Sheridan et al., "Variation in Neural Development as a Result of Exposure to Institutionalization in Early Childhood," *Proceedings of the National Academy of Sciences of the United States of America* 109, no. 32 (2012): 12927–12932, https://doi.org/10.1073/pnas.1200041109.

21. Joseph Castro, "How a Mother's Love Changes a Child's Brain," *Live Science*, January 30, 2012, accessed October 1, 2019, https://www.livescience.com/18196 -maternal-support-child-brain.html.

22. Nim Tottenham and Laurel J. Gabard-Durnam, "The Developing Amygdala: A Student of the World and a Teacher of the Cortex," *Current Opinion in Psychology* 17 (2017): 55–60, https://doi.org/ 10.1016/j.copsyc.2017.06.012.

23. Yolanda R. Chassiakos et al., "Children and Adolescents and Media: Technical Report from the American Academy of Pediatrics," *Pediatrics* 138, no. 5 (2016): e20162593, https://doi.org/10.1542/peds.2016-2593.

24. Dimitri A. Christakis and Frederick J. Zimmerman, "Young Children and Media: Limitations of Current Knowledge and Future Directions of Research," *American Behavioral Scientist* 52, no. 8 (2009): 1177–1185, https://doi.org/10.1177/0002764 209331540.

25. Nielsen, "The Total Audience Report: Q4 2015," March 24, 2016, accessed October 14, 2019, https://www.nielsen.com/us/en/insights/report/2016/the-total-audience-report-q4-2015.

26. Hilda K. Kabali et al., "Exposure and Use of Mobile Media Devices by Young Children," *Pediatrics* 136, no. 6 (2015): 1044–1050, https://doi.org/10.1542/peds.2015-2151.

3. Wired to Connect

1. Carl D. Marci et al., "Physiologic Correlates of Perceived Therapist Empathy and Social-Emotional Process during Psychotherapy," *Journal of Nervous and Mental Disease* 195, no. 2 (2007): 103–111, https://doi.org/10.1097/01.nmd.0000253731.71025.fc.

2. André Felix Gentil et al., "Physiological Responses to Brain Stimulation during Limbic Surgery: Further Evidence of Anterior Cingulate Modulation of Autonomic Arousal," *Biological Psychiatry* 66, no. 7 (2009): 695–701, https://doi.org/10.1016/j.biopsych.2009.05.009.

3. Richard V. Palumbo et al., "Interpersonal Autonomic Physiology: A Systematic Review of the Literature," *Personality and Social Psychology Review* 21, no. 2 (2017): 99–141, https://doi.org/10.1177/1088868316628405.

4. Alejandro Pérez, Manuel Carreiras, and Jon Andoni Duñabeitia, "Brain-to-Brain Entrainment: EEG Interbrain Synchronization while Speaking and Listening," *Scientific Reports* 7, no. 1 (2017), art. 4190, https://doi.org/10.1038/s41598-017-04464-4; "Our Brains Synchronize during a Conversation," *Science Daily,* July 20, 2017, https://www.sciencedaily.com/releases/2017/07/170720095035.htm.

5. Morten L. Kringelbach et al., "A Specific and Rapid Neural Signature for Parental Instinct," *PLoS One* 3, no. 2 (2008): e1664, https://doi.org/10.1371/journal.pone.0001664.

6. Matthew D. Lieberman, *Social: Why Our Brains Are Wired to Connect* (Oxford: Oxford University Press, 2013).

7. Pew Research Center, "Three Technology Revolutions: Internet & Technology," n.d., accessed October 1, 2019, http://www.pewinternet.org/three-technology-revolutions; Steve Hasker, "The World According to Nielsen," presentation at the Festival of Media Conference, April 30, 2014, https://www.youtube.com/watch?v=9O3vETqRW8I.

8. Robin I. M. Dunbar, Anna Marriott, and N. D. C. Duncan, "Human Conversational Behavior," *Human Nature* 8 (1997): 231–246, https://doi.org/10.1007/BF02912493; Mor Naaman, Jeffrey Boase, and Chih-Hui Lai, "Is It Really about Me? Message Content in Social Awareness Streams," *Proceedings of the 2010 ACM Conference on Computer Supported Cooperative Work* (Savannah, GA: Associa-

tion for Computing Machinery, 2010): 189–192, http://infolab.stanford.edu/~mor/research/naamanCSCW10.pdf.

9. Diana I. Tamar and Jason P. Mitchell, "Disclosing Information about the Self Is Intrinsically Rewarding," *Proceedings of the National Academy of Sciences of the United States of America* 109, no. 21 (2012): 8038–8043, https://doi.org/10.1073/pnas.1202129109; Lance Brown, "New Harvard Study Shows Why Social Media Is So Addictive for Many," WTWH Marketing Lab, May 11, 2012, https://marketing.wtwhmedia.com/new-harvard-study-shows-why-social-media-is-so-addictive-for-many/?cn-reloaded=1.

10. "The New Normal: Parents, Teens, and Devices around the World," Common Sense Media, October 1, 2019, https://www.commonsensemedia.org/research/The-New-Normal-Parents-Teens-and-Devices-Around-the-World.

11. Sean Casey, "2016 Nielsen Social Media Report," January 17, 2017, http://www.nielsen.com/us/en/insights/reports/2017/2016-nielsen-social-media-report.html.

12. Jessica Clement, "Number of Monthly Active Facebook Users Worldwide as of 3rd Quarter 2020," *Statista,* accessed September 20, 2021, https://www.statista.com/statistics/264810/number-of-monthly-active-facebook-users-worldwide.

13. "Facebook Reports Fourth Quarter and Full Year 2016 Results," Facebook Investor Relations, February 1, 2017, https://investor.fb.com/investor-news/press-release-details/2017/facebook-Reports-Fourth-Quarter-and-Full-Year-2016-Results/default.aspx; "Facebook's Annual Revenue from 2009 to 2020," *Statista,* February 5, 2021, https://www.statista.com/statistics/268604/annual-revenue-of-facebook.

14. Sonam Sheth, "Russia's Disinformation Campaign on Facebook Could Have Been More Widespread than We Knew," *Business Insider,* September 27, 2017, http://www.businessinsider.com/russia-trolls-facebook-ukraine-activists-disinformation-2017-9; Julia Angwin, Madeleine Varner, and Ariana Tobin, "Facebook Enabled Advertisers to Reach 'Jew Haters,'" *ProPublica,* September 14, 2017, https://www.propublica.org/article/facebook-enabled-advertisers-to-reach-jew-haters?itx[idio]=6119922&ito=792&itq=26975a34-4c8e-47b3-863b-93ce74160069; Kate Irby, "Why Are People Live-Streaming Their Suicides?" *Miami Herald,* January 27, 2017, http://www.miamiherald.com/news/nation-world/national/article129120064.html; Jennifer Valentino-DeVries and Gabriel J. X. Dance, "Facebook Encryption Eyed in Fight against Online Child Sex Abuse," *New York Times,* October 2, 2019.

15. Georgia Wells, Jeff Horwitz, and Deepa Seetharaman, "Facebook Knows Instagram Is Toxic for Teen Girls, Company Documents Show," *Wall Street Journal,* September 14, 2021; Ryan Mac and Sheera Frenkel, "No More Apologies: Inside Facebook's Push to Defend Its Image," *New York Times,* September 21, 2021.

16. James Vincent, "Former Facebook Exec Says Social Media Is Ripping Apart Society," *The Verge,* December 11, 2017, https://www.theverge.com/2017/12/11/16761016/former-facebook-exec-ripping-apart-society; "Chamath Palihapitiya, Founder and CEO Social Capital, on Money as an Instrument of Change," YouTube, November 13, 2017, https://youtu.be/PMotykwoSIk; "The Silicon Valley Sensation: 'Chaos Monkeys'—A *New York Times* Look West Event," YouTube, December 8, 2016, https://www.youtube.com/watch?v=7teIDZXoeZs.

17. Rhett Jones, "Zuckerberg's Campaign Tour 2020 Pulls Over to Explain What a Truck Stop Is," *Gizmodo,* June 24, 2017, https://gizmodo.com/zuckerberg-campaign-tour-2020-pulls-over-to-explain-wha-1796393696.

18. Robin I. M. Dunbar, "Neocortex Size as a Constraint on Group Size in Primates," *Journal of Human Evolution* 22, no. 6 (1992): 469–493, https://doi.org/10.1016/0047-2484(92)90081-J.

19. Robin I. M. Dunbar, "Co-Evolution of Neocortical Size, Group Size and Language in Humans," *Behavioral and Brain Sciences* 16, no. 4 (1993): 681–735, https://doi.org/10.1017/S0140525X00032325.

20. For critique of Dunbar's theory, see, for example, Jan de Ruiter, Gavin Weston, and Stephen M. Lyon, "Dunbar's Number: Group Size and Brain Physiology in Humans Reexamined," *American Anthropologist* 113, no. 4 (2011): 1548–1433, https://doi.org/10.1111/j.1548-1433.2011.01369.x.

21. Joanne L. Powell et al., "Orbital Prefrontal Cortex Volume Correlates with Social Cognitive Competence," *Neuropsychologia* 48, no. 12 (2010): 3554–3562, https://doi.org/10.1016/j.neuropsychologia.2010.08.004.

22. Penelope Lewis et al., "Ventromedial Prefrontal Volume Predicts Understanding of Others and Social Networks Size," *NeuroImage* 57, no. 4 (2011): 1624–1629, https://doi.org/10.1016/j.neuroimage.2011.05.030; Joanne L. Powell et al., "Orbital Prefrontal Cortex Volume Predicts Social Network Size: An Imaging Study of Individual Differences in Humans," *Proceedings of the Royal Society B* 279, no. 1736 (2012): 2157–2162, https://doi.org/10.1098/rspb.2011.2574.

23. Will Knight, "Three Questions for Robin Dunbar: The British Anthropologist's Pioneering Research on Human Social Behavior Has Shaped Business Theory, Military Planning, and Social-Network Design," *MIT Technology Review,* July 12, 2012, https://www.technologyreview.com/s/428478/three-questions-for-robin-dunbar.

4. Early Childhood, Interrupted

1. K. C. Ifeanyi, "How 9-Year-Old YouTube Millionaire Ryan Kaji Is Building a Kid's Media Empire," *Fast Company,* April 23, 2021, https://www.fastcompany.com/90619551/how-9-year-old-youtube-millionaire-ryan-kaji-is-building-a-kids-media-empire

2. On the making of *Sesame Street,* see generally, Michael Davis, *Street Gang: The Complete History of Sesame Street* (New York: Viking Press, 2008).

3. Theresa Treutler, Brian Levine, and Carl D. Marci, "Biometrics and Multi-Platform Messaging: The Medium Matters," *Journal of Advertising Research* 50, no. 3 (2010): 243–249, https://doi.org/10.2501/S0021849910091415; Ellen Rose, "Continuous Partial Attention: Reconsidering the Role of Online Learning in the Age of Interruption," *Educational Technology* 50, no. 4 (2010): 41–46, https://www.jstor.org/stable/i40186198.

4. "Kids Are Gluttons for Digital Video: They Rely Heavily on It for Entertainment," *eMarketer,* December 18, 2017, https://www.emarketer.com/content/kids-are-gluttons-for-digital-video.

5. Yolanda R. Chassiakos et al., "Children and Adolescents and Media: Technical Report from the American Academy of Pediatrics," *Pediatrics* 138, no. 5 (2016): e20162593, https://doi.org/10.1542/peds.2016-2593.

6. Vaughan Bell, Dorothy V. M. Bishop, and Andrew K. Przybylski, "The Debate over Digital Technology and Young People," *BMJ* 351 (2015): h3064, https://doi.org/10.1136/bmj.h3064.

7. C. Walters, "Creator of Baby Einstein Vids Admits in 2005 She Didn't Know What She Was Doing," Consumerist, October 29, 2009, https://consumerist.com/2009/10/29/creator-of-baby-einstein-vids-admitted-in-2005-she-didnt-know-what-she-was-doing.

8. Kate Spencer, "The Facebook Moms: How a Group of Mothers I've Never Even Met Have Helped Me Survive Raising Kids," *Slate,* January 26, 2018, https://slate.com/human-interest/2018/01/the-facebook-moms-group-that-has-helped-me-raise-kids-without-going-crazy.html.

9. Richard Verrier, "Disney Buys Toy Maker, Publisher Baby Einstein," *Los Angeles Times,* November 7, 2001.

10. Frederick J. Zimmerman, Dimitri A. Christakis, and Andrew N. Meltzoff, "Associations between Media Viewing and Language Development in Children under Age 2 Years," *Journal of Pediatrics* 151, no. 4 (2007): 364–368, https://doi.org/10.1016/j.jpeds.2007.04.071; Frederick J. Zimmerman, "Where's the Beef? A Comment on Ferguson and Donnellan (2014)," *Developmental Psychology* 50, no. 1 (2014): 138–140, https://doi.org/10.1037/a0035087.

11. Alice Park, "Baby Einsteins: Not So Smart After All," *Time,* August 6, 2007; Ruth Graham, "The Rise and Fall of Baby Einstein," *Slate,* December 19, 2017, https://slate.com/technology/2017/12/the-rise-and-fall-of-baby-einstein.html.

12. Angeline S. Lillard et al., "Further Examination of the Immediate Impact of Television on Children's Executive Function," *Developmental Psychology* 51, no. 6 (2015): 792–805, http://dx.doi.org/10.1037/a0039097.

13. Daniel R. Anderson and Tiffany A. Pempek, "Television and Very Young Children," *American Behavioral Scientist* 48, no. 5 (2005): 505–522, https://doi.org/10.1177

/0002764204271506; Rachel Barr, "Memory Constraints on Infant Learning from Picture Books, Television, and Touchscreens," *Child Development Perspectives* 7, no. 4 (2013): 205–210, https://doi.org/10.1111/cdep.12041.

14. Judy S. DeLoache et al., "Do Babies Learn from Baby Media?" *Psychological Science* 21, no. 11 (2010): 1570–1574, https://doi.org/10.1177%2F0956797610384145.

15. Dimitri A. Christakis and Frederick J. Zimmerman, "Young Children and Media: Limitations of Current Knowledge and Future Directions of Research," *American Behavioral Scientist* 52, no. 8 (2009): 1177–1185, https://doi.org/10.1177/000276 4209331540.

16. Sarah Roseberry, Kathy Hirsh-Pasek, and Roberta M. Golinkoff, "Skype Me! Socially Contingent Interactions Help Toddlers Learn Language," *Child Development* 85, no. 3 (2014): 956–970, https://doi.org/10.1111/cdev.12166.

17. Dimitri A. Christakis et al., "Early Television Exposure and Subsequent Attentional Problems in Children," *Pediatrics* 113, no. 4 (2004): 708–713, https://doi.org /10.1542/peds.113.4.708; Victoria Clayton, "What's to Blame for the Rise in ADHD? Researchers Point Fingers at TV, Genetics, Overdiagnosis," NBC News.com, September 8, 2009, http://www.nbcnews.com/id/5933775/ns/health-childrens _health/t/whats-blame-rise-adhd.

18. Carlin J. Miller et al., "Brief Report: Television Viewing and Risk for Attention Problems in Preschool Children," *Journal of Pediatric Psychology* 32, no. 4 (2007): 448–452, https://doi.org/10.1093/jpepsy/jsl035.

19. Ine Beyens, Patti M. Valkenburg, and Jessica T. Piotrowski, "Screen Media Use and ADHD-Related Behaviors: Four Decades of Research," *Proceedings of the National Academy of Sciences of the United States of America* 115, no. 40 (2018): 8975–9881, https://doi.org/10.1073/pnas.1611611114.

20. Melissa L. Danielson et al., "Prevalence of Parent-Reported ADHD Diagnosis and Associated Treatment among U.S. Children and Adolescents, 2016," *Journal of Clinical Child and Adolescent Psychology* 47, no. 2 (2018): 199–212, https://doi.org /10.1080/15374416.2017.1417860; "Trends in the Parent-Report of Health Care Provider-Diagnosis and Medication Treatment for ADHD: United States, 2003–2011," Centers for Disease Control and Prevention, n.d., https://www.cdc.gov /ncbddd/adhd/features/key-findings-adhd72013.html; Lisa Rapaport, "More than One in 10 U.S. Kids Have ADHD as Diagnosis Rates Surge," Reuters Health News, December 8, 2015, https://www.reuters.com/article/us-health-adhd -diagnosis-surge/more-than-one-in-10-u-s-kids-have-adhd-as-diagnosis-rates -surge-idUSKBN0TR2SJ20151208.

21. Matthew A. Lapierre, Jessica T. Piotrowski, and Deborah L. Linebarger, "Background Television in the Homes of U.S. Children," *Pediatrics* 130, no. 5 (2012): 839–846, https://doi.org/10.1542/peds.2011-2581; Victoria J. Rideout, "The Common Sense Census: Media Use by Kids Age Zero to Eight," Common Sense Media,

2017, https://www.commonsensemedia.org/research/the-common-sense-census -media-use-by-kids-age-zero-to-eight-2017.

22. Marie Evans Schmidt et al., "The Effects of Background Television on the Toy Play Behavior of Very Young Children," *Child Development* 79, no. 4 (2008): 1137–1151, https://doi.org/10.1111/j.1467-8624.2008.01180.x; Tiffany A. Pempek, Heather L. Kirkorian, and Daniel R. Anderson, "The Effects of Background Television on the Quantity and Quality of Child-Directed Speech by Parents," *Journal of Children and Media* 8, no. 3 (2014): 211–222, https://doi.org/10.1080/17482798.2014.920715.

23. Chassiakos et al., "Children and Adolescents and Digital Media."

24. Roseberry, Hirsh-Pasek, and Golinkoff, "Skype Me!"; Katherine Harmon, "How Important Is Physical Contact with Your Infant?" *Scientific American*, May 6, 2010, https://www.scientificamerican.com/article/infant-touch.

25. Nancy Garon, Susan E. Bryson, and Isabel M. Smith, "Executive Function in Pre- schoolers: A Review Using an Integrative Framework," *Psychological Review* 134 (2008): 31–60, https://doi.org/10.1037/0033-2909.134.1.31.

26. Dimitri A. Christakis et al., "Modifying Media Content for Preschool Children: A Randomized Controlled Trial," *Pediatrics* 131, no. 3 (2013): 431–438, https://dx .doi.org/10.1542%2Fpeds.2012-1493; Rachel Barr et al., "Infant and Early Child- hood Exposure to Adult-Directed and Child-Directed Television Programming: Relations with Cognitive Skills at Age Four," *Merrill-Palmer Quarterly* 56, no. 1 (2010): 21–48, https://doi.org/10.1353/mpq.0.0038.

27. Patricia M. Greenfield, "Technology and Informal Education: What Is Taught, What Is Learned," *Science* 323, no. 5910 (2009): 69–71, https://doi.org/10.1126/sci- ence.1167190; Sarah Vaala, Anna Ly, and Michael H. Levine, "Getting a Read on the App Stores: A Market Scan and Analysis of Children's Literacy Apps," The Joan Ganz Cooney Center at Sesame Workshop, Fall 2015, www.joanganz cooneycenter.org/wp-content/uploads/2015/12/jgcc_gettingaread.pdf.

28. Alexis R. Lauricella, Rachel Barr, and Sandra L. Calvert, "Parent-Child Interac- tions during Traditional and Computer Storybook Reading for Children's Com- prehension: Implications for Electronic Storybook Design," *International Journal of Child-Computer Interaction* 2, no. 1 (2014): 17–25, https://doi.org/10.1016/j .ijcci.2014.07.001; Gabrielle A. Strouse, Katherine O'Doherty, and Georgene L. Troseth, "Effective Coviewing: Preschoolers' Learning from Video after a Dia- logic Questioning Intervention," *Developmental Psychology* 49, no. 12 (2013): 2368–2382, https://doi.org/10.1037/a0032463.

29. Ellen E. Wartella et al., "Parenting in the Age of Digital Technology: A National Survey," Report of the Center on Media and Human Development, School of Communications, Northwestern University, June 2014, https://cmhd.north western.edu/wp-content/uploads/2015/06/ParentingAgeDigitalTechnology .REVISED.FINAL_.2014.pdf.

30. Alexis Hiniker et al., "Screen Time Tantrums: How Families Manage Screen Media Experiences for Toddlers and Preschoolers," *Proceedings of the 2016 CHI Conference on Human Factors in Computing Systems* (New York: Association for Computing Machinery, 2016): 648–660, https://dl.acm.org/doi/pdf/10.1145/2858036.2858278.

31. Kathy Hirsh-Pasek et al., "Putting Education in 'Educational' Apps: Lessons from the Science of Learning," *Psychological Science in the Public Interest* 16, no. 1 (2015): 3–34, https://doi.org/10.1177%2F1529100615569721.

32. Stanislas Dehaene and Laurent Cohen, "Cultural Recycling of Cortical Maps," *Neuron* 56, no. 2 (2007): 384–398, https://doi.org/10.1016/j.neuron.2007.10.004; Eric I. Knudsen, "Sensitive Periods in the Development of the Brain and Behavior," *Journal of Cognitive Neuroscience* 16, no. 8 (2004): 1412–1425, https://doi.org/10.1162/0898929042304796.

33. Adriana G. Bus and Marinus H. van Ijzendoorn, "Affective Dimension of Mother-Infant Picturebook Reading," *Journal of School Psychology* 35, no. 1 (1997): 47–60, https://doi.org/10.1016/S0022-4405(96)00030-1; Christopher J. Lonigan and Timothy Shanahan, "Developing Early Literacy: Report of the National Early Literacy Panel," National Institute for Literacy, January 2009, https://lincs.ed.gov/publications/pdf/NELPReport09.pdf; Tzipi Horowitz-Kraus and John S. Hutton, "Brain Connectivity in Children Is Increased by the Time They Spend Reading Books and Decreased by the Length of Exposure to Screen-Based Media," *Acta Paediatrica* 107, no. 4 (2018): 685–693, https://doi.org/10.1111/apa.14176; Pamela C. High and Perri Klass, "Literacy Promotion: An Essential Component of Primary Care Pediatric Practice," *Pediatrics* 134, no. 2 (2014): 404–409, https://doi.org/10.1542/peds.2014-1384.

34. Michael J. Kieffer, Rose K. Vukovic, and Daniel Berry, "Roles of Attention Shifting and Inhibitory Control in Fourth-Grade Reading Comprehension, *Reading Research Quarterly* 48, no. 4 (2013): 333–348, https://doi.org/10.1002/rrq.54; Micaela E. Christopher et al., "Predicting Word Reading and Comprehension with Executive Function and Speed Measures across Development: A Latent Variable Analysis," *Journal of Experimental Psychology: General* 141, no. 3 (2012): 470–488, https://dx.doi.org/10.1037%2Fa0027375; Tzipi Horowitz-Kraus, "Can the Error-Monitoring System Differentiate ADHD from ADHD with Reading Disability? Reading and Executive Dysfunction as Reflected in Error Monitoring," *Journal of Attention Disorders* 20, no. 10 (2016): 889–902, https://doi.org/10.1177%2F1087054713488440; Tzipi Horowitz-Kraus et al., "Increased Resting-State Functional Connectivity of Visual- and Cognitive-Control Brain Networks after Training in Children with Reading Difficulties," *NeuroImage: Clinical* 8 (2015): 619–630, https://dx.doi.org/10.1016%2Fj.nicl.2015.06.010.

35. "Finding Their Story," Kids & Family Reading Report, 7th ed., 2017, Scholastic, https://www.scholastic.com/content/dam/KFRR/Downloads/KFRReport_Finding%20Their%20Story.pdf.

36. Yunqi Zhu, Hong Zhang, and Mei Tian, "Molecular and Functional Imaging of Internet Addiction," *BioMed Research International* 2015 (2015): 378675, https://doi.org/10.1155/2015/378675; Soon-Beom Hong et al., "Decreased Functional Brain Connectivity in Adolescents with Internet Addiction," *PLoS One* 8, no. 2 (2013): e57831, https://doi.org/10.1371/journal.pone.0057831; Hikaru Takeuchi et al., "The Impact of Television Viewing on Brain Structures: Cross-Sectional and Longitudinal Analyses," *Cerebral Cortex* 25, no. 5 (2015): 1188–1197, https://doi.org/10.1093/cercor/bht315.

37. John S. Hutton et al., "Shared Reading Quality and Brain Activation during Story Listening in Preschool Age Children," *Journal of Pediatrics* 191 (2017): 204–211, https://doi.org/10.1016/j.jpeds.2017.08.037.

38. Horowitz-Kraus and Hutton, "Brain Connectivity in Children."

39. Adriana G. Bus, Zsofia K. Takacs, and Cornelia A. Kegel, "Affordances and Limitations of Electronic Storybooks for Young Children's Emergent Literacy," *Developmental Review* 35 (2015): 79–97, https://doi.org/10.1016/j.dr.2014.12.004.

40. Stephanie Reich, Joanna Yau, and Mark Warschauer, "Tablet-Based eBooks for Young Children: What Does the Research Say?" *Journal of Developmental and Behavioral Pediatrics* 37, no. 7 (2016): 585–591, https://doi.org/10.1097/DBP.0000000000000335; "Toddlers Need Laps More than Apps," CBS News video clip, December 9, 2018, https://www.cbsnews.com/video/toddlers-need-laps-more-than-apps.

41. Rideout, "The Common Sense Census: Media Use by Kids Age Zero to Eight"; Victoria Rideout, Ulla G. Foehr, and Donald F. Roberts, "Generation M²: Media in the Lives of 8–18 Year-Olds," Henry J. Kaiser Family Foundation, January 2010, https://www.kff.org/wp-content/uploads/2013/04/8010.pdf.

42. Rideout, "The Common Sense Census: Media Use by Kids Age Zero to Eight."

43. Annie M. Paul, "Your Wired Kid," *Good Housekeeping*, March 6, 2012, http://www.goodhousekeeping.com/life/parenting/tips/a19226/children-overuse-electronics-technology; Rideout, "The Common Sense Census: Media Use by Kids Age Zero to Eight."

44. Rideout, Foehr, and Roberts, "Generation M²," 2010; Rideout, "The Common Sense Census: Media Use by Kids Age Zero to Eight"; Anastasia Kononova, "Multitasking across Borders: A Cross-National Study of Media Multitasking Behaviors, Its Antecedents, and Outcomes," *International Journal of Communications* 7, no. 23 (2013): 1688–1710, https://ijoc.org/index.php/ijoc/article/view/2119.

45. Radesky and Christakis, "Increased Screen Time."

46. Melina R. Uncapher et al., "Media Multitasking and Cognitive, Psychological, Neural and Learning Differences," *Pediatrics* 140, Suppl. 2 (2017): S62–66, https://doi.org/ 10.1542/peds.2016-1758D.

47. Winneke A. van der Schuur et al., "The Consequences of Media Multitasking for Youth: A Review," *Computers in Human Behavior* 53 (2015): 204–215, https://doi .org/10.1016/j.chb.2015.06.035.

48. Terrie E. Moffitt et al., "A Gradient of Childhood Self-Control Predicts Health, Wealth, and Public Safety," *Proceedings of the National Academy of Sciences of the United States of America* 108, no. 7 (2011): 2693–2698, http://dx.doi.org/10.1073 /pnas.1010076108.

49. Angela L. Duckworth, "The Significance of Self-Control," *Proceedings of the National Academy of Sciences of the United States of America* 108, no. 7 (2011): 2639–2640, https://dx.doi.org/10.1073%2Fpnas.1019725108.

50. Ellie Dolgin, "The Myopia Boom: Short-Sightedness Is Reaching Epidemic Proportions," *Nature News,* March 18, 2015, http://www.nature.com/news/the-myopia -boom-1.17120.

51. Seang-Mei Saw et al., "Component Dependent Risk Factors for Ocular Parameters in Singapore Chinese Children," *Ophthalmology* 109, no. 11 (2002): 2065–2071, https://doi.org/10.1016/S0161-6420(02)01220-4.

52. Pei-Chang Wu et al., "Outdoor Activity during Class Recess Reduces Myopia Onset and Progression in School Children," *Ophthalmology* 120, no. 5 (2013): 1080–1085, https://doi.org/10.1016/j.ophtha.2012.11.009.

53. Marita Feldkaemper and Frank Schaeffel, "An Updated View on the Role of Dopamine in Myopia," *Experimental Eye Research* 114 (2013): 106–119, https://doi.org /10.1016/j.exer.2013.02.007.

54. Cheryl D. Fryar, Margaret D. Carroll, and Cynthia L. Ogden, "Prevalence of Overweight and Obesity among Children and Adolescents: United States, 1963–1965 through 2011–2012," Centers for Disease Control and Prevention, Health E-Stats, September 2014, https://www.cdc.gov/nchs/data/hestat/obesity_child_11 _12/obesity_child_11_12.htm.

55. Thomas N. Robinson et al., "Screen Media Exposure and Obesity in Children and Adolescents," *Pediatrics* 140, Suppl. 2 (2017): S97–S101, https://doi.org/10.1542/ peds.2016-1758K; William H. Dietz and Steven L. Gortmaker, "Do We Fatten Our Children at the Television Set? Obesity and Television Viewing in Children and Adolescents," *Pediatrics* 75, no. 5 (1985): 807–812, https://pediatrics.aap publications.org/content/75/5/807; Robert J. Hancox, Barry J. Milne, and Richie Poulton, "Association between Child and Adolescent Television Viewing and Adult Health: A Longitudinal Birth Cohort Study," *Lancet* 364, no. 9430 (2004): 257–262, https://doi.org/10.1016/S0140-6736(04)16675-0.

56. Thomas N. Robinson, "Reducing Children's Television Viewing to Prevent Obesity: A Randomized Controlled Trial," *Journal of the American Medical Association* 282, no. 16 (1999): 1561–1567, https://doi.org/10.1001/jama.282.16.1561; Leonard H. Epstein et al., "A Randomized Trial of the Effects of Reducing Television Viewing and Computer Use on Body Mass Index in Young Children," *Archives of Pediatric and Adolescent Medicine* 162, no. 3 (2008): 239–245, https://doi.org/10.1001/archpediatrics.2007.45.

57. Natalie Pearson and Stuart J. Biddle, "Sedentary Behavior and Dietary Intake in Children, Adolescents, and Adults: A Systematic Review," *American Journal of Preventive Medicine* 41, no. 2 (2011): 178–188, https://doi.org/10.1016/j.amepre.2011.05.002.

58. Erik Näslund and Per M. Hellstrom, "Appetite Signaling: From Gut Peptides and Enteric Nerves to the Brain," *Physiology and Behavior* 92, no. 1–2 (2007): 256–262, https://doi.org/10.1016/j.physbeh.2007.05.017.

59. Jennifer L. Harris, John A. Bargh, and Kelly D. Brownell, "Priming Effects of Television Food Advertising on Eating Behavior," *Health Psychology* 28, no. 4 (2009): 404–413, https://doi.org/10.1037/a0014399.

60. "The Impact of Food Advertising on Childhood Obesity," American Psychological Association, 2010, https://www.apa.org/topics/kids-media/food.

5. Tweens, Teens, and Tech

1. Graham L. Baum et al., "Modular Segregation of Structural Brain Networks Supports the Development of Executive Function in Youth," *Current Biology* 27, no. 11 (2017): 1561–1572.E8, https://doi.org/10.1016/j.cub.2017.04.051; "Brain Images Reveal Roots of Kids' Increasing Cognitive Control," *Science Daily,* May 25, 2017, https://www.sciencedaily.com/releases/2017/05/170525123048.htm.

2. Russell A. Poldrack, "Can Cognitive Processes Be Inferred from Neuroimaging Data?" *Trends in Cognitive Science* 10, no. 2 (2006): 59–63, https://doi.org/10.1016/j.tics.2005.12.004.

3. Jay N. Giedd, "The Amazing Teen Brain," *Scientific American Mind,* May 2016, https://www.scientificamerican.com/article/the-amazing-teen-brain.

4. Baum et al., "Modular Segregation."

5. Victoria Rideout and Michael B. Robb, "The Common Sense Census: Media Use by Tweens and Teens," Common Sense Media, 2019, https://www.commonsensemedia.org/research/the-common-sense-census-media-use-by-tweens-and-teens-2019; Monica Anderson and Jingjing Jiang, "Teens, Social Media and Technology 2018," Pew Research Center, May 31, 2018, https://www.pewinternet.org/2018/05/31/teens-social-media-technology-2018.

6. Alex Cocotas, "Chart of the Day: Kids Send a Mind Boggling Number of Texts Every Month," *Business Insider*, March 22, 2013, http://www.businessinsider.com/chart-of-the-day-number-of-texts-sent-2013-3; Amanda Lenhart, "Teens, Smartphones and Texting," Pew Research Center, March 19, 2012, http://www.pewinternet.org/2012/03/19/teens-smartphones-texting.

7. Megan A. Moreno et al., "Internet Use and Multitasking among Older Adolescents: An Experience Sampling Approach," *Computers in Human Behavior* 28, no. 4 (2012): 1097–1102, https://doi.org/10.1016/j.chb.2012.01.016.

8. Rosalina Richards et al., "Adolescent Screen Time and Attachment to Parents and Peers," *Archives of Pediatric and Adolescent Medicine* 164, no. 3 (2010): 258–262, https://doi.org/10.1001/archpediatrics.2009.280.

9. Kelly M. Lister-Landman, Sarah E. Domoff, and Eric F. Dubow, "The Role of Compulsive Texting in Adolescents' Academic Functioning," *Psychology of Popular Media Culture* 6, no. 4 (2017): 311–325, http://dx.doi.org/10.1037/ppm0000100.

10. Qin Chen, "The Hidden Costs of Letting Your Children Be Raised by Screens and Smart Devices," CNBC, February 22, 2018, https://www.cnbc.com/2018/02/22/the-hidden-costs-of-letting-your-children-be-raised-by-screens-and-smart-devices.html.

11. Lauren E. Sherman et al., "The Power of the Like in Adolescence: Effects of Peer Influence on Neural and Behavioral Responses to Social Media," *Psychological Science* 27, no. 7 (2016): 1027–1035, https://dx.doi.org/10.1177%2F0956797616645673.

12. Stuart Wolpert, "The Teenage Brain on Social Media: The Findings of a New UCLA Study Shed Light on the Influences of Peers and Much More," UCLA Newsroom, May 31, 2016, http://newsroom.ucla.edu/releases/the-teenage-brain-on-social-media

13. Laurel J. Felt and Michael B. Robb, "Technology Addiction: Concern, Controversy, and Finding a Balance," Common Sense Media, San Francisco, 2016, https://www.commonsensemedia.org/research/technology-addiction-concern-controversy-and-finding-balance; Lenhart, "Teens, Smartphones and Texting"; Monica Anderson and Jingjing Jiang, "Teens' Social Media Habits and Experiences," Pew Research Center, Internet and Technology, November 28, 2018, http://www.pewinternet.org/2018/11/28/teens-social-media-habits-and-experiences.

14. Kate Davis, "Young People's Digital Lives: The Impact of Interpersonal Relationships and Digital Media Use on Adolescents' Sense of Identity," *Computers in Human Behavior* 29 (2013): 2281–2293, https://doi.org/10.1016/j.chb.2013.05.022; Emily Weinstein, "The Social Media See-Saw: Positive and Negative Influences on Adolescents' Affective Well-Being," *New Media and Society* 20, no. 10 (2018): 3597–3623, https://doi.org/10.1177/1461444818755634.

15. John R. Tanner et al., "How Business Students Spend Their Time—Do They Really Know?" *Proceedings of the Academy of Educational Leadership* 13, no. 2 (2008): 81–85; Muhammet Demirbilek and Tarik Talan, "The Effect of Social Media Multitasking on Classroom Performance," *Active Learning in Higher Education* 19, no. 2 (2018): 117–129, https://doi.org/10.1177%2F1469787417721382.

16. Joan O'C. Hamilton, "Spoiling Our Kids," *Stanford Magazine,* November / December 2002, https://stanfordmag.org/contents/spoiling-our-kids.

17. Donald F. Roberts, Ulla G. Foehr, and Victoria Rideout, "Generation M: Media in the Lives of 8–18 Year-Olds," Henry J. Kaiser Family Foundation, March 2005, https://www.kff.org/wp-content/uploads/2013/01/generation-m-media-in-the-lives-of-8-18-year-olds-report.pdf.

18. Victoria Rideout, Ulla G. Foehr, and Donald F. Roberts, "Generation M^2: Media in the Lives of 8-to-18 Year-Olds," Henry J. Kaiser Family Foundation, January 2010, https://www.kff.org/wp-content/uploads/2013/04/8010.pdf; Roberts, "Generation M," 2005; Victoria J. Rideout, "The Common Sense Census: Media Use by Kids Age Zero to Eight," Common Sense Media, 2017, https://www.commonsensemedia.org/research/the-common-sense-census-media-use-by-kids-age-zero-to-eight-2017.

19. Hagit Magen, "The Relations between Executive Functions, Media Multitasking and Polychronicity," *Computers in Human Behavior* 67 (2017): 1–9, https://doi.org/10.1016/j.chb.2016.10.011.

20. Mona Moisala et al., "Media Multitasking Is Associated with Distractibility and Increased Prefrontal Activity in Adolescents and Young Adults," *NeuroImage* 134 (2016): 113–121, https://doi.org/10.1016/j.neuroimage.2016.04.011.

21. Melina R. Uncapher et al., "Media Multitasking and Cognitive, Psychological, Neural and Learning Differences," *Pediatrics* 140, Suppl. 2 (2017): S62–S66, https://doi.org/10.1542/peds.2016-1758D.

22. Jean M. Twenge, "Have Smartphones Destroyed a Generation?" *Atlantic,* September 2017.

23. Lloyd D. Johnston et al., *Monitoring the Future: National Survey Results on Drug Use, 1975–2017,* "Key Findings on Adolescent Drug Use, 2017," Institute for Social Research, University of Michigan, Ann Arbor, January 2018, http://www.monitoringthefuture.org//pubs/monographs/mtf-overview2017.pdf; Twenge, "Have Smartphones Destroyed a Generation?"

24. Andrew K. Przybylski and Netta Weinstein, "A Large-Scale Test of the Goldilocks Hypothesis: Quantifying the Relations between Digital-Screen Use and the Mental Well-Being of Adolescents," *Psychological Science* 28, no. 2 (2017): 204–215, https://doi.org/10.1177/0956797616678438; "Teens Unlikely to Be Harmed by Moderate Digital Screen Use," *Science Daily,* January 13, 2017, www.sciencedaily.com/releases/2017/01/170113085940.htm; Dimitri A. Christakis, "The Challenges

of Defining and Studying 'Digital Addiction' in Children," *JAMA* 321, no. 23 (2019): 2277–2278, https://doi.org/10.1001/jama.2019.4690.

25. Ramin Mojtabai, Mark Olfsen, and Beth Han, "National Trends in the Prevalence and Treatment of Depression in Adolescents and Young Adults," *Pediatrics* 138, no. 6 (2016): e20161878, https://doi.org/10.1542/peds.2016-1878; Randy P. Auerbach et al., "WHO World Mental Health Surveys International College Study Project: Prevalence and Distribution of Mental Disorders," *Journal of Abnormal Psychology* 127, no. 7 (2018): 623–638, https://www.apa.org/pubs/journals/releases/abn-abn0000362.pdf; Henry A. Spiller et al., "Sex- and Age-Specific Increases in Suicide Attempts by Self-Poisoning in the United States among Youth and Young Adults from 2000 to 2018," *Journal of Pediatrics* 210 (2019): 201–208, https://doi.org/10.1016/j.jpeds.2019.02.045.

26. Mingli Liu, Lang Wu, and Shuqiao Yao, "Dose–Response Association of Screen Time-Based Sedentary Behaviour in Children and Adolescents and Depression: A Meta-Analysis of Observational Studies," *British Journal of Sports Medicine* 50, no. 20 (2016): 1252–1258, http://dx.doi.org/10.1136/bjsports-2015-095084; Melissa G. Hunt et al., "No More FOMO: Limiting Social Media Decreases Loneliness and Depression," *Journal of Social and Clinical Psychology* 37, no. 10 (2018): 751–768, https://doi.org/10.1521/jscp.2018.37.10.751.

27. "Social Media Use Increases Depression and Loneliness, Study Finds," *Science Daily,* November 8, 2018, https://www.sciencedaily.com/releases/2018/11/181108164316.htm.

28. Susanna Schrobsdorff, "There's a Startling Increase in Major Depression among Teens in the U.S.," *Time,* November 16, 2016.

29. Frank Bruni, "Today's Exhausted Superkids," *New York Times,* July 29, 2015.

30. Rebecca H. Bitsko et al., "Epidemiology and Impact of Health Care Provider–Diagnosed Anxiety and Depression among US Children," *Journal of Developmental and Behavioral Pediatrics* 39, no. 5 (2018): 395–403, https://dx.doi.org/10.1097%2FDBP.0000000000000571; Robin Wilson, "An Epidemic of Anguish: Overwhelmed by Demand for Mental-Health Care, Colleges Face Conflicts in Choosing How to Respond," *Chronicle of Higher Education,* August 31, 2015, https://www.chronicle.com/article/An-Epidemic-of-Anguish/232721.

31. Katherine M. Keyes et al., "The Great Sleep Recession: Changes in Sleep Duration among US Adolescents 1991–2012," *Pediatrics* 135, no. 3 (2015): 460–468, https://doi.org/10.1542/peds.2014-2707; Judith Owens et al., "Insufficient Sleep in Adolescents and Young Adults: An Update on Causes and Consequences," *Pediatrics* 134, no. 3 (2014): e921–e932, https://doi.org/10.1542/peds.2014-1696.

32. Christina J. Calamaro, Thornton B. Mason, and Sarah J. Ratcliffe, "Adolescents Living the 24/7 Lifestyle: Effects of Caffeine and Technology on Sleep Duration and Daytime Functioning," *Pediatrics* 123, no. 6 (2009): e1005–1010, https://doi

.org/10.1542/peds.2008-3641; Lauren Hale and Stanford Guan, "Screen Time and Sleep among School-Aged Children and Adolescents: A Systematic Literature Review," *Sleep Medicine Reviews* 21 (2015): 50–58, https://doi.org/10.1016/j.smrv .2014.07.007.

33. Mary A. Carskadon, Cecelia Vieira, and Christine Acebo, "Association between Puberty and Delayed Phase Preference," *Sleep* 16, no. 3 (1993): 258–262, https://doi .org/10.1093/sleep/16.3.258.

34. Stephanie J. Crowley and Mary A. Carskadon, "Modifications to Weekend Recovery Sleep Delay Circadian Phase in Older Adolescents," *Chronobiology International* 27, no. 7 (2010): 1469–1492, https://doi.org/10.3109/07420528.2010.503293.

35. Rebecca L. Orbeta et al., "High Caffeine Intake in Adolescents: Associations with Difficulty Sleeping and Feeling Tired in the Morning," *Journal of Adolescent Health* 38, no. 4 (2006): 451–453, https://doi.org/10.1016/j.jadohealth.2005.05.014; Bjorn Rasch and Jan Born, "About Sleep's Role in Memory," *Psychological Reviews* 93, no. 2 (2013): 681–766, https://doi.org/10.1152/physrev.00032.2012.

36. Eric Suni, "How Much Sleep Do We Really Need?" National Sleep Foundation, updated March 10, 2021, https://www.sleepfoundation.org/how-sleep-works/how -much-sleep-do-we-really-need; Ray C. Meldrum and Emily Restivo, "The Behavioral and Health Consequences of Sleep Deprivation among U.S. High School Students: Relative Deprivation Matters," *Preventive Medicine* 63 (2014): 24–28, https://doi.org/10.1016/j.ypmed.2014.03.006; Owens et al., "Insufficient Sleep in Adolescents and Young Adults"; Amy R. Wolfson and Mary A. Carskadon, "Understanding Adolescents' Sleep Patterns and School Performance: A Critical Appraisal," *Sleep Medicine Reviews* 7, no. 6 (2003): 491–506, https://doi.org/10.1016/ S1087-0792(03)90003-7; Jennifer A. O'Dea, Michael J. Dibley, and N. M. Rankin, "Low Sleep and Low Socioeconomic Status Predict High Body Mass Index: A 4-Year Longitudinal Study of Australian Schoolchildren," *Pediatric Obesity* 7, no. 4 (2012): 295–303, https://doi.org/10.1111/j.2047-6310.2012.00054.x; Nicola Spiers et al., "Age and Birth Cohort Differences in the Prevalence of Common Mental Disorder in England: National Psychiatric Morbidity Surveys 1993–2007," *British Journal of Psychiatry* 198, no. 6 (2011): 479–484, https://doi.org/10.1192/bjp. bp.110.084269; Dean W. Beebe, "Cognitive, Behavioral, and Functional Consequences of Inadequate Sleep in Children and Adolescents," *Pediatric Clinics of North America* 58, no. 3 (2011): 649–665, https://doi.org/10.1016/j.pcl.2011.03.002.

37. James E. Gangwisch et al., "Earlier Parental Set Bedtimes as a Protective Factor against Depression and Suicidal Ideation," *Sleep* 33, no. 1 (2010): 97–106, https://dx .doi.org/10.1093%2Fsleep%2F33.1.97.

38. Brian Y. Park et al., "Is Internet Pornography Causing Sexual Dysfunctions? A Review with Clinical Reports," *Behavioral Sciences* 6, no. 3 (2016): 17, https://dx .doi.org/10.3390%2Fbs6030017.

39. Simone Kühn and Jürgan Gallinat, "Brain Structure and Functional Connectivity Associated with Pornography Consumption: The Brain on Porn," *JAMA Psychiatry* 71, no. 7 (2014): 827–834, https://doi.org/10.1001/jamapsychiatry.2014.93.

40. Jon E. Grant, Judson A. Brewer, and Marc N. Potenza, "The Neurobiology of Substance and Behavioral Addictions," *CNS Spectrums* 11, no. 12 (2006): 189–206, https://doi.org/10.1017/S109285290001511X; Rita Z. Goldstein and Nora D. Volkow, "Dysfunction of the Prefrontal Cortex in Addiction: Neuroimaging Findings and Clinical Implications," *Nature Reviews Neuroscience* 12 (2011): 652–669, https://doi.org/10.1038/nrn3119; Todd Love et al., "Neuroscience of Internet Pornography Addiction: A Review and Update," *Behavioral Sciences* 5, no. 3 (2015): 388–433, https://doi.org/10.3390/bs5030388.

6. Adults, There Will Be Consequences

1. "The Nielsen Total Audience Report: Q2 2017," Nielsen Company, November 16, 2017, https://www.nielsen.com/us/en/insights/reports/2017/the-nielsen-total-audience-q2-2017.html; "U.S. Consumers Are Shifting the Time They Spend with Media," Nielsen Company, March 19, 2019, https://www.nielsen.com/us/en/insights/news/2019/us-consumers-are-shifting-the-time-they-spend-with-media.html.

2. Deirdre Barrett, *Supernormal Stimuli: How Primal Urges Overran Their Evolutionary Purpose* (New York: W. W. Norton, 2010).

3. See, generally, Nikolaas Tinbergen, *The Study of Instinct* (Oxford: Oxford University Press, 1951).

4. Darryl T. Gwynne and David C. F. Rentz, "Beetles on the Bottle: Male Buprestids Mistake Stubbies for Females (Coleoptera)," *Australian Journal of Entomology* 22, no. 1 (1983): 79–80, https://doi.org/10.1111/j.1440-6055.1983.tb01846.x.

5. T. N. C. Vidya, "Supernormal Stimuli and Responses," *Resonance* 23, no. 8 (2018): 853–860, https://doi.org/10.1007/s12045-018-0688-x.

6. Barbara Demick, "Gamers Rack Up Losses," *Los Angeles Times,* August 29, 2005.

7. Cecilia Cheng and Angel Yee-Iam Li, "Internet Addiction Prevalence and Quality of (Real) Life: A Meta-Analysis of 31 Nations across Seven World Regions," *Cyberpsychology, Behavior, and Social Networking* 17, no. 12 (2014): 755–760, https://dx.doi.org/10.1089%2Fcyber.2014.0317; Cecilia Cheng et al., "Prevalence of Social Media Addiction across 32 Nations: Meta-Analysis with Sub-Group Classification Schemes and Cultural Values," *Addictive Behaviors* 117 (2021): 106845, https://doi.org/10.1016/j.addbeh.2021.106845.

8. American Psychiatric Association, "DSM-5 Fact Sheets: Internet Gaming Disorder," American Psychiatric Association Educational Resources, 2013, https://

www.psychiatry.org/psychiatrists/practice/dsm/educational-resources/dsm-5
-fact-sheets.

9. Todd Love et al., "Neuroscience of Internet Pornography Addiction: A Review
   and Update," *Behavioral Sciences* 5, no. 3 (2015): 388–433, https://doi.org/10.3390/
   bs5030388; American Psychiatric Association, "DSM-5 Fact Sheets: Internet
   Gaming Disorder."

10. Chih-Hung Ko et al., "Proposed Diagnostic Criteria of Internet Addiction for
    Adolescents," *Journal of Nervous and Mental Disease* 193, no. 11 (2005): 728–733,
    https://doi.org/10.1097/01.nmd.0000185891.13719.54.

11. Andrew K. Przybylski, Netta Weinstein, and Kou Murayama, "Internet Gaming
    Disorder: Investigating the Clinical Relevance of a New Phenomenon," *Amer-
    ican Journal of Psychiatry* 171, no. 3 (2017): 230–236, https://doi.org/10.1176/appi
    .ajp.2016.16020224.

12. Marianne Littel et al., "Electrophysiological Indices of Biased Cognitive Pro-
    cessing of Substance-Related Cues: A Meta-Analysis," *Neuroscience and Biobe-
    havioral Reviews* 36, no. 8 (2012): 1803–1816, https://doi.org/10.1016/j.neubiorev
    .2012.05.001.

13. Wolfram Schultz, "Reward Prediction Error," *Current Biology* 27 (2017): R369–R271,
    https://doi.org/10.1016/j.cub.2017.02.064.

14. M. Victoria Puig, Evan G. Antzoulatos, and Earl K. Miller, "Prefrontal Dopamine
    in Associative Learning and Memory," *Neuroscience* 282 (2014): 217–229, https://dx
    .doi.org/10.1016%2Fj.neuroscience.2014.09.026.

15. Ashley N. Gearhardt et al., "Neural Correlates of Food Addiction," *Archives of
    General Psychiatry* 68, no. 8 (2011): 808–816, https://doi.org/10.1001/archgen
    psychiatry.2011.32.

16. Gaetano Di Chiara et al., "Homologies and Differences in the Action of Drugs
    of Abuse and a Conventional Reinforcer (Food) on Dopamine Transmission: An
    Interpretive Framework of the Mechanism of Drug Dependence," *Advances in
    Pharmacology* 42 (1997): 983–987, https://doi.org/10.1016/S1054-3589(08)60911-4;
    Christopher M. Olsen, "Natural Rewards, Neuroplasticity, and Non-Drug Ad-
    dictions," *Neuropharmacology* 61, no. 7 (2011): 1109–1122, https://dx.doi.org/10
    .1016%2Fj.neuropharm.2011.03.010.

17. Nora D. Volkow et al., "Addiction: Decreased Reward Sensitivity and Increased
    Expectation Sensitivity Conspire to Overwhelm the Brain's Control Circuit,"
    Bioessays 32, no. 9 (2010): 748–755, https://dx.doi.org/10.1002/bies.201000042.

18. Raymundo Báez-Mendoza and Wolfram Schultz, "The Role of the Striatum in
    Social Behavior," *Frontiers in Neuroscience* 7 (2013), art. 233, https://dx.doi.org/10
    .3389%2Ffnins.2013.00233.

19. Ann M. Graybiel and Kyle S. Smith, "Good Habits, Bad Habits: Researchers Are
    Pinpointing the Brain Circuits That Can Help Us Form Good Habits and

Break Bad Ones," *Scientific American* 310, no. 6 (2014): 38–43, https://10.1038 /scientificamerican0614-38.

20. Lee N. Robins, "The Vietnam Drug User Returns," Special Action Office for Drug Abuse Prevention monograph, series A, no. 2 (Washington, DC: US Government Printing Office, May 1974); Lee N. Robins et al., "Vietnam Veterans Three Years after Vietnam: How Our Study Changed Our View of Heroin," 1977, reprinted in *American Journal on Addictions* 19, no. 3 (2010): 203–211, https://doi.org/10.1111/ j.1521-0391.2010.00046.x.

21. Alix Spiegel, "What Vietnam Taught Us about Breaking Bad Habits," National Public Radio, January 2, 2012, https://www.npr.org/sections/health-shots/2012/01 /02/144431794/what-vietnam-taught-us-about-breaking-bad-habits.

22. Jon E. Grant et al., "Introduction to Behavioral Addictions," *American Journal of Drug Alcohol Abuse* 36, no. 5 (2010): 233–241, https://doi.org/10.3109/00952990 .2010.491884.

23. Betsy Sparrow, Jenny Liu, and Daniel M. Wegner, "Google Effects on Memory: Cognitive Consequences of Having Information at Our Fingertips," *Science* 333, no. 6043 (2011): 776–778, https://doi.org/10.1126/science.1207745.

24. Daniel M. Wegner and Adrian F. Ward, "How Google Is Changing Your Brain," *Scientific American* 309, no. 6 (2013): 58–61, https://doi.org/10.1038/scientific american1213-58.

25. Nash Unsworth et al., "Working Memory and Fluid Intelligence: Capacity, Attention Control, and Secondary Memory Retrieval," *Cognitive Psychology* 71 (2014): 1–26, https://doi.org/10.1016/j.cogpsych.2014.01.003.

26. Alan D. Baddeley and Graham Hitch, "Working Memory," in *The Psychology of Learning and Motivation*, vol. 8, ed. Gordon H. Bower, 47–89 (New York: Academic Press, 1974).

27. Edward K. Vogel and Maro G. Machizawa, "Neural Activity Predicts Individual Differences in Visual Working Memory Capacity," *Nature* 428 (2004): 748–751, https://doi.org/10.1038/nature02447.

28. John M. Gaspar et al., "Inability to Suppress Salient Distractors Predicts Low Visual Working Memory Capacity," *Proceedings of the National Academy of Sciences of the United States of America* 113, no. 13 (2016): 3693–3698, https://doi.org /10.1073/pnas.1523471113.

29. Andrew B. Leber, "Neural Predictors of Within-Subject Fluctuations in Attention Control," *Journal of Neuroscience* 30, no. 34 (2010): 11458–11465, https://doi .org/10.1523/JNEUROSCI.0809-10.2010.

30. Ellen Rose, "Continuous Partial Attention: Reconsidering the Role of Online Learning in the Age of Disruption," *Educational Technology* 50, no. 4 (2010): 41–46, https://www.jstor.org/stable/i40186198.

31. William Yardley, "Clifford Nass, Who Warned of a Data Deluge, Dies at 55," *New York Times,* November 6, 2013.

32. Eyal Ophir, Clifford Nass, and Anthony D. Wagner, "Cognitive Control in Media Multitaskers," *Proceedings of the National Academy of Sciences of the United States of America* 106, no. 37 (2009): 15583–15587, https://doi.org/10.1073/pnas.0903620106; Adam Gorlick, "Media Multitaskers Pay Mental Price, Stanford Study Shows," *Stanford News,* August 24, 2009, https://news.stanford.edu/2009/08/24/multitask -research-study-082409.

33. "Multitasking: Switching Costs," American Psychological Association, March 20, 2006, https://www.apa.org/research/action/multitask.

34. Caterina Rechichi, Gilda De Mojà, and Pasquale Aragona, "Video Game Vision Syndrome: A New Clinical Picture in Children?" *Journal of Pediatric Ophthalmology and Strabismus* 54, no. 6 (2017): 346–355, https://doi.org/10.3928/01913913 -20170510-01.

35. M. Logaraj, V. Madhupriya, and Shailendra K. Hegde, "Computer Vision Syndrome and Associated Factors among Medical and Engineering Students in Chennai," *Annals of Medical and Health Sciences Research* 4, no. 2 (2014): 179–185, https://doi.org/10.4103/2141-9248.129028; T. R. Akinbinu and Y. J. Mashalla, "Impact of Computer Technology on Health: Computer Vision Syndrome (CVS)," *Medical Practice and Review* 5, no. 3 (2014): 20–30, https://doi.org/10.5897 /MPR.2014.0121.

36. Akinbinu and Mashalla, "Impact of Computer Technology"; Sowjanya Gowrisankaran et al., "Asthenopia and Blink Rate under Visual and Cognitive Loads," *Optometry and Vision Science* 89, no. 1 (2012): 97–104, https://doi.org/10.1097 /OPX.0b013e318236dd88.

37. Ewa Gustafsson et al., "Texting on Mobile Phones and Musculoskeletal Disorders in Young Adults: A Five-Year Cohort Study," *Applied Ergonomics* 58 (2017): 208–214, https://doi.org/10.1016/j.apergo.2016.06.012.

38. Kenneth K. Hansraj, "Assessment of Stresses in the Cervical Spine Caused by Posture and Position of the Head," *Surgical Technology International* 25 (2014): 277–279, https://www.ncbi.nlm.nih.gov/pubmed/25393825.

39. Lindsey Bever, "'Text Neck' Is Becoming an 'Epidemic' and Could Wreck Your Spine," *Washington Post,* November 20, 2014.

40. Eva Blozik et al., "Depression and Anxiety as Major Determinants of Neck Pain: A Cross-Sectional Study in General Practice," *BMC Musculoskeletal Disorders* 10 (2009): 13–21, https://doi.org/10.1186/1471-2474-10-13.

7. Impact on Mental Health

1. Elias Aboujaoude et al., "Cyberbullying: Review of an Old Problem Gone Viral," *Journal of Adolescent Health* 57, no. 1 (2015): 10–18, https://doi.org/10.1016/j. jadohealth.2015.04.011.

2. Doriana Chialant, Judith Ebersheim, and Bruce H. Price, "The Dialectic between Empathy and Violence: An Opportunity for Intervention?" *Journal of Neuropsychiatry and Clinical Neurosciences* 28, no. 4 (2016): 273–285, https://doi.org/10.1176/appi.neuropsych.15080207.

3. Bradley C. Taber-Thomas et al., "Arrested Development: Early Prefrontal Lesions Impair the Maturation of Moral Judgement," *Brain* 137, no. 4 (2014): 1254–1261, https://doi.org/10.1093/brain/awt377.

4. Sara H. Konrath, Edward H. O'Brian, and Courtney Hsing, "Changes in Dispositional Empathy in American College Students over Time: A Meta-Analyses," *Personality and Social Psychology Review* 15, no. 2 (2011): 180–198, https://doi.org/10.1177%2F1088868310377395.

5. Shalini Misra et al., "The iPhone Effect: The Quality of In-Person Social Interactions in the Presence of Mobile Devices," *Environment and Behavior* 48, no. 2 (2014): 1–24, https://doi.org/10.1177/0013916514539755.

6. Andrew K. Przybylski and Netta Weinstein, "Can You Connect with Me Now? How the Presence of Mobile Communication Technology Influences Face-to-Face Conversation Quality," *Journal of Social and Personal Relationships* 30, no. 3 (2013): 237–246, https://doi.org/10.1177/0265407512453827.

7. Sherry Turkle, "Stop Googling, Let's Talk," *New York Times,* September 26, 2015.

8. Adrian F. Ward et al., "Brain Drain: The Mere Presence of One's Own Smartphone Reduces Available Cognitive Capacity," *Journal of the Association of Consumer Research* 2, no. 2 (2017): 140–154, http://dx.doi.org/10.1086/691462.

9. Jillian H. Fecteau and Douglas P. Munoz, "Salience, Relevance, and Firing: A Priority Map for Target Selection," *Trends in Cognitive Sciences* 10, no. 8 (2006): 382–390, https://doi.org/10.1016/j.tics.2006.06.011.

10. Samuel Evans et al., "Getting the Cocktail Party Started: Masking Effects in Speech Perception," *Journal of Cognitive Neuroscience* 28, no. 3 (2016): 483–500, https://doi.org/10.1162/jocn_a_00913.

11. William A. Johnston and Veronica J. Dark, "Selective Attention," *Annual Review of Psychology* 37 (1986): 43–75, https://doi.org/10.1146/annurev.ps.37.020186.000355.

12. "Americans Say They Are More Anxious than a Year Ago; Baby Boomers Report Greatest Increase in Anxiety," news release, American Psychological Association, May 7, 2018, https://www.psychiatry.org/newsroom/news-releases/americans-say-they-are-more-anxious-than-a-year-ago-baby-boomers-report-greatest-increase-in-anxiety; "Any Anxiety Disorder," National Institute of Mental Health, November 2017, https://www.nimh.nih.gov/health/statistics/any-anxiety-disorder.shtml#part_155096; Borwin Bandelow and Sophie Michaelis, "Epidemiology of Anxiety Disorders in the 21st Century," *Dialogues in Clinical Neuroscience* 17, no. 3 (2015): 327–335, https://doi.org/10.31887/DCNS.2015.17.3/bbandelow.

13. "The High Price of Silence: Analyzing the Business Implications of an Under-Vacationed Workforce," U.S. Travel Association, October 12, 2016, https://www.ustravel.org/sites/default/files/media_root/document/High_Price-of_Silence_FINAL.pdf.

14. Alex Williams, "Prozac Nation Is Now the United States of Xanax," *New York Times,* June 10, 2017.

15. Jennifer Garam, "Social Media Makes Me Feel Bad about Myself: Reading Facebook and Twitter Streams Can Destroy My Self-Esteem," *Psychology Today* blog, September 26, 2011, https://www.psychologytoday.com/us/blog/progress-not-perfection/201109/social-media-makes-me-feel-bad-about-myself.

16. Robert Kraut et al., "Internet Paradox: A Social Technology That Reduces Social Involvement and Psychological Well-Being?" *American Psychologist* 52, no. 9 (1998): 1017–1031, https://doi.org/10.1037//0003-066X.53.9.1017.

17. Robert Kraut et al., "The Internet Paradox Revisited," *Journal of Social Issues* 58, no. 1 (2002): 49–74, https://doi.org/10.1111/1540-4560.00248.

18. Holly B. Shakya and Nicholas A. Christakis, "Association of Facebook Use with Compromised Well-Being: A Longitudinal Study," *American Journal of Epidemiology* 185, no. 3 (2017): 203–211, https://doi.org/10.1093/aje/kww189.

19. Ruoyun Lin and Sonja Utz, "The Emotional Responses of Browsing Facebook: Happiness, Envy, and the Role of Tie Strength," *Computers in Human Behavior* 52 (2015): 29–38, https://doi.org/10.1016/j.chb.2015.04.064.

20. Brian A. Primack et al., "Use of Multiple Social Media Platforms and Symptoms of Depression and Anxiety: A Nationally-Representative Study among U.S. Young Adults," *Computers in Human Behavior* 69 (2017): 1–9, https://doi.org/10.1016/j.chb.2016.11.013.

21. Joanna Davila, "Skills for Healthy Romantic Relationships," TEDxSBU, November 17, 2015, https://www.youtube.com/watch?v=gh5VhaicC6g.

22. Joanne Davila et al., "Frequency and Quality of Social Networking among Young Adults: Associations with Depressive Symptoms, Rumination, and Corumination," *Psychology of Popular Media Culture* 1, no. 2 (2012): 72–86, https://doi.org/10.1037/a0027512.

23. On rumination and co-rumination, see Jason S. Spendelow, Laura M. Simmonds, and Rachel E. Avery, "The Relationship between Co-Rumination and Internalizing Problems: A Systematic Review and Meta-Analysis," *Clinical Psychology and Psychotherapy* 24 (2017): 512–527, https://doi.org/10.1002/cpp.2023; Amanda J. Rose, Wendy Carlson, and Erika M. Waller, "Prospective Associations of Co-Rumination with Friendship and Emotional Adjustment: Considering the Socioemotional Trade-Offs of Co-Rumination," *Developmental Psychology* 43, no. 4 (2007): 1019–1031, https://doi.org/10.1037/0012-1649.43.4.1019.

24. Davila et al., "Frequency and Quality of Social Networking."

25. Daniel P. Johnson and Mark A. Whisman, "Gender Differences in Rumination: A Meta-Analysis," *Personality and Individual Differences* 55, no. 4 (2013): 367–374, https://doi.org/10.1016/j.paid.2013.03.019.

26. Yingkai Yang et al., "The Relationships between Rumination and Core Executive Functions: A Meta-Analysis," *Depression and Anxiety* 34, no. 1 (2017): 37–50, https://doi.org/10.1002/da.22539.

27. Ian H. Gotlib and Jutta Joormann, "Cognition and Depression: Current Status and Future Directions," *Annual Review of Clinical Psychology* 6 (2010): 285–312, https://doi.org/10.1146/annurev.clinpsy.121208.131305; Leanne M. Williams, "Defining Biotypes for Depression and Anxiety Based on Large-Scale Circuit Dysfunction: A Theoretical Review of the Evidence and Future Directions for Clinical Translation," *Depression and Anxiety* 34 (2017): 9–24, https://doi.org/10.1002/da.22556; Greg J. Siegle et al., "Can't Shake That Feeling: fMRI Assessment of Sustained Amygdala Activity in Response to Emotional Information in Depressed Individuals," *Biological Psychiatry* 51, no. 9 (2002): 693–707, https://doi.org/10.1016/S0006-3223(02)01314-8.

28. Greg J. Siegle, Cameron S. Carter, and Michael E. Thase, "Use of fMRI to Predict Recovery from Unipolar Depression with Cognitive Behavior Therapy," *American Journal of Psychiatry* 163, no. 4 (2006): 735–738, https://doi.org/10.1176/appi.ajp.163.4.735.

29. Samantha S. Rosenthal et al., "Negative Experiences on Facebook and Depressive Symptoms in Young Adults," *Journal of Adolescent Health* 59, no. 5 (2016): 510–516, https://doi.org/10.1016/j.jadohealth.2016.06.023.

30. "2019 National Survey of Drug Use and Health Annual National Report," National Institute of Mental Health, September 11, 2020, https://www.samhsa.gov/data/report/2019-nsduh-annual-national-report; World Health Organization, "'Depression: Let's Talk' Says WHO, as Depression Tops List of Causes of Ill Health," WHO news release, March 30, 2017, https://www.who.int/news/item/30-03-2017--depression-let-s-talk-says-who-as-depression-tops-list-of-causes-of-ill-health.

31. Andrea H. Weinberger et al., "Trends in Depression Prevalence in the USA from 2005 to 2015: Widening Disparities in Vulnerable Groups," *Psychological Medicine* 48, no. 8 (2018): 1308–1315, https://doi.org/10.1017/S0033291717002781.

32. World Health Organization, "'Depression: Let's Talk' Says WHO."

33. Peter Wehrwein, "Astounding Increase in Antidepressant Use by Americans," Harvard Health blog, October 20, 2011, https://www.health.harvard.edu/blog/astounding-increase-in-antidepressant-use-by-americans-201110203624.

34. Cristiano Lima, "A Whistleblower's Power: Key Takeaways from the Facebook Papers," *Washington Post*, October 26, 2021; Jeff Horwitz and Deepa Seetharaman, "Facebook Executives Shut Down Efforts to Make the Site Less Divisive," *Wall Street Journal*, May 26, 2020.

35. R. W. B. Lewis, *Edith Wharton: A Biography* (New York: Harper and Row, 1975); Leon Festinger, "A Theory of Social Comparison Processes," *Human Relations* 7, no. 2 (1954), https://doi.org/10.1177/001872675400700202.

36. Sun Young Park and Young Min Baek, "Two Faces of Social Comparison on Facebook: The Interplay between Social Comparison Orientation, Emotions, and Psychological Well-Being," *Computers in Human Behavior* 79 (2018): 83–93, https://doi.org/10.1016/j.chb.2017.10.028.

37. Paul Gilbert et al., "Relationship of Anhedonia and Anxiety to Social Rank, Defeat, and Entrapment," *Journal of Affective Disorders* 71, no. 1–3 (2002): 141–151, https://doi.org/10.1016/S0165-0327(01)00392-5; Steven Allan and Paul Gilbert, "A Social Comparison Scale: Psychometric Properties and Relationship to Psychopathology," *Personality and Individual Differences* 19, no. 3 (1995): 293–299, https://doi.org/10.1016/0191-8869(95)00086-L; Judith B. White et al., "Frequent Social Comparisons and Destructive Emotions and Behaviors: The Dark Side of Social Comparisons," *Journal of Adult Development* 13, no. 1 (2006): 36–44, https://doi.org/10.1007/s10804-006-9005-0.

38. Park and Baek, "Two Faces of Social Comparison on Facebook."

39. Justin W. Moyer, "When Facebook Friends Become Depressing," *Washington Post*, April 7, 2015.

40. Mai-Ly N. Steers, Robert E. Wickham, and Linda K. Acitelli, "Seeing Everyone Else's Highlight Reels: How Facebook Usage Is Linked to Depressive Symptoms," *Journal of Social and Clinical Psychology* 33, no. 8 (2014): 701–731, https://doi.org/10.1521/jscp.2014.33.8.701.

41. Garam, "Social Media Makes Me Feel Bad."

42. Eve Caligor, Kenneth N. Levy, and Frank E. Yeomans, "Narcissistic Personality Disorder: Diagnostic and Clinical Challenges," *American Journal of Psychiatry* 172, no. 5 (2015): 415–422, https://doi.org/10.1176/appi.ajp.2014.14060723.

43. Laura E. Buffardi and W. Keith Campbell, "Narcissism and Social Networking Websites," *Personality and Social Psychology Bulletin* 34, no. 10 (2008): 1303–1324, https://doi.org/10.1177/0146167208320061.

44. Jean M. Twenge, "Egos Inflating over Time: A Cross-Temporal Meta-Analysis of the Narcissistic Personality Inventory," *Journal of Personality* 76, no. 4 (2008): 875–902, https://doi.org/10.1111/j.1467-6494.2008.00507.x.

45. Eunike Wetzel et al., "The Narcissism Epidemic Is Dead; Long Live the Narcissism Epidemic," *Psychological Science* 28, no. 12 (2017): 1833–1847, https://doi.org/10.1177/0956797617724208.

46. Timo Gnambs and Markus Appel, "Narcissism and Social Networking Behavior: A Meta-Analysis," *Journal of Personality* 86, no. 2 (2018): 200–212, https://doi.org/10.1111/jopy.12305.

47. Brittany Gentile et al., "The Effect of Social Networking Websites on Positive Self-Views: An Experimental Investigation," *Computers in Human Behavior* 28, no. 5 (2012): 1929–1933, https://doi.org/10.1016/j.chb.2012.05.012.

48. Ravi Chandra, "Is Facebook Making Us Narcissistic?" *Psychology Today* blog, February 5, 2018, https://www.psychologytoday.com/us/blog/the-pacific-heart /201802/is-facebook-making-us-narcissistic; W. Keith Campbell and Jean M. Twenge, "Narcissism Unleashed," Association for Psychological Science *Observer* 26, no. 10 (December 2013): 28–29, https://www.psychologicalscience.org/observer /narcissism-unleashed.

49. Caligor, Levy, and Yeomans, "Narcissistic Personality Disorder."

50. Lars Schulze et al., "Gray Matter Abnormalities in Patients with Narcissistic Personality Disorder," *Journal of Psychiatric Research* 47, no. 10 (2013): 1363–1369, https://doi.org/10.1016/j.jpsychires.2013.05.017.

51. Claus Lamm, Jean Decety, and Tania Singer, "Meta-Analytic Evidence for Common and Distinct Neural Networks Associated with Directly Experienced Pain and Empathy for Pain," *NeuroImage* 54, no. 3 (2011): 2492–2502, https://doi .org/10.1016/j.neuroimage.2010.10.014.

52. Adam Waytz and Kurt Gray, "Does Online Technology Make Us More or Less Sociable? A Preliminary Review and Call for Research," *Perspectives on Psychological Science* 13, no. 4 (2018): 473–491, https://doi.org/10.1177/1745691617746509.

53. John T. Cacioppo and William Patrick, *Loneliness: Human Nature and the Need for Connection* (New York: W. W. Norton, 2008).

54. George Masnick, "The Rise of the Single-Person Household," Housing Perspectives, Harvard Joint Center for Housing Studies, May 20, 2015, https://www.jchs .harvard.edu/blog/the-rise-of-the-single-person-household.

55. Susie Demarinis, "Loneliness at Epidemic Levels in America," *Explore* 16, no. 5 (2020): 278–279, https://doi.org/10.1016/j.explore.2020.06.008; "Former Surgeon General Sounds the Alarm on the Loneliness Epidemic," *CBS This Morning*, October 19, 2017, https://www.cbsnews.com/news/loneliness-epidemic-former -surgeon-general-dr-vivek-murthy.

56. Leah D. Doane and Emma K. Adam, "Loneliness and Cortisol: Momentary, Day-to-Day and Trait Associations," *Psychoneuroendocrinology* 35, no. 3 (2010): 430–441, https://dx.doi.org/10.1016%2Fj.psyneuen.2009.08.005.

57. Julianne Holt-Lunstad, Timothy B. Smith, and J. Bradley Layton, "Social Relationships and Mortality Risk: A Meta-Analytic Review, *PLoS Medicine,* July 27, 2010, https://doi.org/10.1371/journal.pmed.1000316.

58. Maureen Ryan, "TV Peaks Again in 2016: Could It Hit 500 Shows in 2017?" *Variety,* December 21, 2016, http://variety.com/2016/tv/news/peak-tv-2016-scripted -tv-programs-1201944237.

59. Carl Marci, "Storytelling in the Digital Media Age," *TechCrunch,* 2015, http://
techcrunch.com/2015/03/02/storytelling-in-the-digital-media-age; Katherine E.
Powers et al., "Social Connection Modulates Perceptions of Animacy," *Psychological Science* 25, no. 10 (2014): 1943–1948, https://doi.org/10.1177/0956797614
547706.

8. First, Recognize the Problem

1. Chih-Hung Ko et al., "Predictive Values of Psychiatric Symptoms for Internet
Addiction in Adolescents: A 2-Year Prospective Study," *Archives of Pediatric
and Adolescent Medicine* 163, no. 10 (2009): 937–943, https://doi.org/10.1001
/archpediatrics.2009.159.

2. Wendy Wood and Dennis Rünger, "Psychology of Habit," *Annual Review of
Psychology* 67 (2016): 289–314, https://doi.org/10.1146/annurev-psych-122414
-033417; Martin Oscarsson et al., "A Large-Scale Experiment on New Year's Resolutions: Approach-Oriented Goals Are More Successful than Avoidance-
Oriented Goals," *PLoS One* 15, no. 12 (2020): e0234097, https://doi.org/10.1371
/journal.pone.0234097.

3. Hengchen Dai, Katherine L. Milkman, and Jason Riis, "The Fresh Start Effect:
Temporal Landmarks Motivate Aspirational Behavior," *Management Science* 60,
no. 10 (2014): 2563–2582.

4. Mark Muraven and Roy F. Baumeister, "Self-Regulation and Depletion of Limited
Resources: Does Self-Control Resemble a Muscle?" *Psychological Bulletin* 126,
no. 2 (2000): 247–259, https://doi.org/10.1037/0033-2909.126.2.247.

5. Judith A. Ouellette and Wendy Wood, "Habit and Intention in Everyday Life:
The Multiple Processes by Which Past Behavior Predicts Future Behavior," *Psychological Bulletin* 124, no. 1 (1998): 54–74, https://doi.org/10.1037/0033-2909
.124.1.54.

6. Matthias Brand, Kimberly S. Young, and Christian Laier, "Prefrontal Control and
Internet Addiction: A Theoretical Model and Review of Neuropsychological and
Neuroimaging Findings," *Frontiers in Human Neuroscience* 8 (2014), art. 375,
https://doi.org/10.3389/fnhum.2014.00375.

7. David T. Neal, Wendy Wood, and Aimee Drolet, "How Do People Adhere to
Goals When Willpower Is Low? The Profits (and Pitfalls) of Strong Habits,"
Journal of Personality and Social Psychology 104, no. 6 (2013): 959–975, https://doi
.org/10.1037/a0032626.

8. National Jewish Health, "Smoking Cessation," *U.S. News & World Report* health
website, n.d., accessed December 31, 2019, https://health.usnews.com/health
-conditions/allergy-asthma-respiratory/smoking-cessation/overview.

9. Katherine L. Milkman, Julia A. Minson, and Kevin G. M. Volpp, "Holding Hunger Games Hostage at the Gym: An Evaluation of Temptation Bundling," *Management Science* 60, no. 2 (2014): 283–299, https://doi.org/10.1287/mnsc.2013.1784.

10. Steven Kotler, *The Rise of Superman: Decoding the Science of Ultimate Human Performance* (New York: New Harvest, 2014), 116–117.

9. Ten Rules for a Healthy Tech-Life Balance

1. Shamsi Iqbal and Eric Horvitz, "Disruption and Recovery of Computing Tasks: Field Study, Analysis, and Directions," *Proceedings of the SIGCHI Conference on Human Factors in Computing Systems,* April 2007: 677–686, https://doi.org/10.1145/1240624.1240730.

2. Laura Vanderkam, "Monotasking Is the New Multitasking," *Fast Company,* August 6, 2013, https://www.fastcompany.com/3015251/monotasking-is-the-new-multitasking.

3. Claire A. Wolniewicza et al., "Problematic Smartphone Use and Relations with Negative Affect, Fear of Missing Out, and Fear of Negative and Positive Evaluation," *Psychiatry Research* 262 (2018): 618–623, https://doi.org/10.1016/j.psychres.2017.09.058.

4. David P. Jarmolowicz et al., "Robust Relation between Temporal Discounting Rates and Body Mass," *Appetite* 78 (2014): 63–67, https://doi.org/10.1016/j.appet.2014.02.013; Natalia Albein-Urios et al., "Monetary Delay Discounting in Gambling and Cocaine Dependence with Personality Comorbidities," *Addictive Behaviors* 39 (2014): 1658–1662, https://doi.org/10.1016/j.addbeh.2014.06.001; James MacKillop et al., "Delayed Reward Discounting and Addictive Behavior: A Meta-Analysis," *Psychopharmacology* 216 (2011): 305–321, https://doi.org/10.1007/s00213-011-2229-0; David J. Hardisty and Elke U. Weber, "Discounting Future Green: Money versus the Environment," *Journal of Experimental Psychology: General* 138 (2009): 329–340, https://doi.org/10.1037/a0016433.

5. Karolina M. Lempert and Elizabeth A. Phelps, "Review: The Malleability of Intertemporal Choice," *Trends in Cognitive Sciences* 20, no. 1 (2016): 64–74, https://doi.org/10.1016/j.tics.2015.09.005; on framing effects, see generally, Daniel Kahneman and Amos Tversky, *Choices, Values, and Frames* (New York: Cambridge University Press, 2000).

6. Nancy M. Petry et al., "Shortened Time Horizons and Insensitivity to Future Consequences in Heroin Addicts," *Addiction* 93 (1998): 729–738, https://doi.org/10.1046/j.1360-0443.1998.9357298.x.

7. David DeSteno et al., "Gratitude: A Tool for Reducing Economic Impatience," *Psychological Science* 25 (2014): 1262–1267, https://doi.org/10.1177/0956797614529979.

8. DeSteno et al., "Gratitude"; Elizabeth A. Olson et al., "White Matter Integrity Predicts Delay Discounting Behavior in 9- to 23-Year-Olds: A Diffusion Tensor Imaging Study," *Journal of Cognitive Neuroscience* 21, no. 7 (2009): 1406–1421, https://doi.org/10.1162/jocn.2009.21107.

9. Erving Goffman, *The Presentation of Self in Everyday Life* (New York: Anchor Books, Doubleday, 1956), esp. 208–237.

10. On identity and the prefrontal cortex, see Paolo Bozzatello et al., "Autobiographical Memories, Identity Disturbance and Brain Functioning in Patients with Borderline Personality Disorder: An fMRI Study," *Heliyon* 5, no. 3 (2019): e01323, https://doi.org/10.1016/j.heliyon.2019.e01323.

11. Joe Otterson, "Delayed Viewing Ratings: 'Roseanne' Returns to the Top," *Variety,* May 21, 2018, https://variety.com/2018/tv/news/delayed-viewing-ratings-roseanne-2-1202817911.

12. John Koblin, "After Racist Tweet, Roseanne Barr's Show Is Cancelled by ABC," *New York Times,* May 29, 2018; Nick Visser, "Roseanne Barr Breaks Down in First Interview since Scandal," *Huffington Post,* June 25, 2018, https://www.huffingtonpost.com/entry/roseanne-barr-apology-podcast_us_5b3049d0e4b0321a01d29c5b.

13. Jennifer Brooks, "Students Remind Us: T.H.I.N.K. before You Post Online," *Star Tribune,* June 2, 2018, http://www.startribune.com/students-remind-us-t-h-i-n-k-before-you-post-online/484390801.

14. James S. House, Karl R. Landis, and Debra Umberson, "Social Relationships and Health," *Science* 241 (1988): 540–545, https://doi.org/10.1126/science.3399889; Julianne Holt-Lunstad, Timothy B. Smith, and J. Bradley Layton, "Social Relationships and Mortality Risk: A Meta-Analytic Review," *PLoS Medicine,* July 27, 2010, https://doi.org/10.1371/journal.pmed.1000316.

15. George M. Slavich and Steven W. Cole, "The Emerging Field of Human Social Genomics," *Clinical Psychological Science* 1, no. 3 (2013): 331–348, https://doi.org/10.1177/2167702613478594.

16. Robert Waldinger, "What Makes a Good Life? Lessons from the Longest Study on Happiness," TEDxBeaconStreet, November 2015, https://www.ted.com/talks/robert_waldinger_what_makes_a_good_life_lessons_from_the_longest_study_on_happiness.

17. Robert J. Waldinger et al., "Security of Attachment to Spouses in Late Life: Concurrent and Prospective Links with Cognitive and Emotional Well-Being," *Clinical Psychological Science* 3, no. 4 (2015): 516–529, https://doi.org/10.1177/2167702614541261; George E. Vaillant et al., "Antecedents of Intact Cognition and Dementia at Age 90 Years: A Prospective Study," *International Journal of Geriatric Psychiatry* 29, no. 12 (2014): 1278–1285, https://doi.org/10.1002/gps.4108.

18. Isabela Granic, Adam Lobel, and Rutger Engels, "The Benefits of Playing Video Games," *American Psychologist* 69, no. 1 (2014): 66–78, https://doi.org/10.1037/a0034857.

19. Helen Riess et al., "Empathy Training for Resident Physicians: A Randomized Controlled Trial of a Neuroscience-Informed Curriculum," *Journal of General Internal Medicine* 27, no. 10 (2012): 1280–1286, https://doi.org/10.1007/s11606-012-2063-z.

20. Yalda T. Uhls et al., "Five Days at Outdoor Education Camp without Screens Improves Preteen Skills with Nonverbal Emotion Cues," *Computers in Human Behavior* 39 (2014): 387–392, https://doi.org/10.1016/j.chb.2014.05.036.

21. Zachary J. Ward et al., "Projected U.S. State-Level Prevalence of Adult Obesity and Severe Obesity," *New England Journal of Medicine* 381 (2019): 2440–2450, https://doi.org/10.1056/NEJMsa1909301; Vanessa Harrar et al., "Food's Visually-Perceived Fat Content Affects Discrimination Speed in an Orthogonal Spatial Task," *Experimental Brain Research* 214, no. 3 (2011): 351–356, https://doi.org/10.1007/s00221-011-2833-6.

22. Charles Spence et al., "Eating with Our Eyes: From Visual Hunger to Digital Satiation," *Brain and Cognition* 110 (2016): 53–63, https://doi.org/10.1016/j.bandc.2015.08.006.

23. Eric Stice et al., "Relation of Reward from Food Intake and Anticipated Food Intake to Obesity: A Functional Magnetic Resonance Imaging Study," *Journal of Abnormal Psychology* 117, no. 4 (2008): 924–935, https://doi.org/10.1037/a0013600.

24. Christian Benedict et al., "Acute Sleep Deprivation Enhances the Brain's Response to Hedonic Food Stimuli: An fMRI Study," *Journal of Clinical Endocrinology and Metabolism* 97, no. 3 (2012): E443–E447, https://doi.org/10.1210/jc.2011-2759.

25. Matthias Brand, Kimberly S. Young, and Christian Laier, "Prefrontal Control and Internet Addiction: A Theoretical Model and Review of Neuropsychological and Neuroimaging Findings," *Frontiers in Human Neuroscience* 8 (2014): 1–13, https://doi.org/10.3389/fnhum.2014.00375.

26. "A Magazine Is an iPad That Does Not Work," October 6, 2011, https://www.youtube.com/watch?v=aXV-yaFmQNk&feature=youtu.be.

27. Ferris Jabr, "The Reading Brain in the Digital Age: Why Paper Still Beats Screens," *Scientific American,* November 2013, https://www.scientificamerican.com/article/the-reading-brain-in-the-digital-age-why-paper-still-beats-screens.

28. Ziming Liu, "Digital Reading: An Overview," *Chinese Journal of Library and Information Science (English Edition)* (2012): 85–94, https://pdfs.semanticscholar.org/4746/aeeb8b966f8ba4520715ba8c3856bacc9c39.pdf.

29. Marlon M. Maducdoc et al., "Visual Consequences of Electronic Reader Use: A Pilot Study," *International Ophthalmology* 37, no. 2 (2017): 433–439, https://doi.org/10.1007/s10792-016-0281-9; Roger Dooley, "Print vs. Digital: Another Emotional

Win for Paper," Neuromarketing blog, June 2015, http://www.neuroscience
marketing.com/blog/articles/print-vs-digital.htm; "Enhancing the Value of Mail:
The Human Response," RARC report no. RARC-WP-15-012, Office of the In-
spector General, U.S. Postal Service, June 15, 2015, https://www.uspsoig.gov
/sites/default/files/document-library-files/2015/rarc-wp-15-012.pdf.

30. Susanne Diekelmann and Jan Born, "The Memory Function of Sleep," *Nature Reviews Neuroscience* 11, no. 2 (2010): 114–126, https://doi.org/10.1038/nrn2762.

31. Chiari Cirelli and Giulino Tononi, "Is Sleep Essential?" *PLoS Biology* 6, no. 8
(2008): 1605–1611, https://doi.org/10.1371/journal.pbio.0060216.

32. Denise J. Cai et al., "REM, Not Incubation, Improves Creativity by Priming As-
sociative Networks," *Proceedings of the National Academy of Sciences of the United
States of America* 106, no. 25 (2009): 10130–10134, https://doi.org/10.1073/pnas
.0900271106; Louisa Lyon, "Is an Epidemic of Sleeplessness Increasing the Incidence of Alzheimer's Disease?" *Brain* 142, no. 6 (2019): e30, https://doi.org
/10.1093/brain/awz087.

33. Ferris Jabr, "Blue LEDs Light Up Your Brain," *Scientific American,* November 1,
2016, https://www.scientificamerican.com/article/blue-leds-light-up-your-brain.

34. Tatsiana Palavets and Mark Rosenfield, "Blue-Blocking Filters and Digital Eye-
strain," *Optometry and Vision Science* 96, no. 1 (2019): 48–54, https://doi.org
/10.1097/OPX.0000000000001318; Mark Rosenfield, "A Double-Blind Test of
Blue-Blocking Filters on Symptoms of Digital Eye Strain," *Work* 65, no. 2 (2020):
343–348, https://doi.org/10.3233/WOR-203086; "Do Blue Light Blocking Glasses
Really Work?" NPR, February 21, 2021, https://www.npr.org/2021/02/21/969886124
/do-blue-light-blocking-glasses-really-work.

35. Bill Loomis, "1900–1930: The Years of Driving Dangerously," *Detroit News,*
April 26, 2015; Rachel Ross, "Who Invented the Traffic Light?" *Live Science,* December 15, 2016, https://www.livescience.com/57231-who-invented-the-traffic
-light.html.

36. National Safety Council, "2018 Marks Third Straight Year That Motor Vehicle
Deaths Are Estimated to Have Reached 40,000," National Safety Council Newsroom, February 13, 2019, https://www.nsc.org/in-the-newsroom/2018-marks
-third-straight-year-that-motor-vehicle-deaths-are-estimated-to-have-reached
-40-000.

37. Neal E. Boudette, "U.S. Traffic Deaths Rise for a Second Straight Year," *New York
Times,* February 17, 2017.

38. Cameron Jahn, "100 Times Worse than We Thought: Insights from Zendrive's
2018 Distracted Driving Snapshot," Zendrive Data Studies, April 10, 2018,
http://blog.zendrive.com/distracted-driving-is-far-worse; "New Data from Cambridge Mobile Telematics Shows Distracted Driving Dangers," Cambridge Mobile Telematics, April 3, 2017, https://www.cmtelematics.com/news/new-data

-cambridge-mobile-telematics-shows-distracted-driving-dangers; Mary Wisniewski, "1 in 4 Drivers Was on a Phone Just before Crash, Study of Distracted Driving Says," *Chicago Tribune,* April 3, 2017.

39. Zendrive, "Zendrive's 2019 Distracted Driving Study," April 2019, https://go1.zendrive.com/distracted-driving-study-2019.

40. Ben Spencer, "Texting while Driving 'Slows Reaction Times More than Drink or Drugs,'" *Daily Mail,* June 8, 2014, http://www.dailymail.co.uk/news/article-2652015/Texting-driving-slows-reaction-times-drink-drugs.html.

41. Thomas A. Dingus et al., "Driver Crash Risk Factors and Prevalence Evaluation Using Naturalistic Driving Data," *Proceedings of the National Academy of Science of the United States of America* 113, no. 10 (2016): 2636–2641, https://doi.org/10.1073/pnas.1513271113.

42. Joel Cooper, Hailey Ingebretsen, and David L. Stayer, "Mental Workload of Common Voice-Based Vehicle Interactions across Six Different Vehicle Systems," AAA Foundation for Traffic Safety, October 2014, https://aaafoundation.org/wp-content/uploads/2018/01/MentalWorkloadofCommonVoiceReport.pdf; Frank A. Drews, Monisha Pasupathi, and David L. Strayer, "Passenger and Cell Phone Conversations in Simulated Driving," *Journal of Experimental Psychology: Applied* 14, no. 4 (2008): 392–400, https://doi.org/10.1037/a0013119.

43. Zendrive, "Zendrive's 2019 Distracted Driving Study"; Neale Martin, "Why TED Talks Don't Change Your Life Much," TEDxPeachtree, November 30, 2013, https://www.youtube.com/watch?v=AHoVGxNrzH4.

44. Chris Teague, "These Age Brackets Admit to the Most Distracted Driving," Digital Trends, August 15, 2019, https://www.digitaltrends.com/cars/distracted-driving-boomers-millennials.

45. Martin, "Why TED Talks Don't Change Your Life Much."

46. Gretchen Rubin, "Stop Expecting to Change Your Habit in 21 Days," *Psychology Today* blog, October 21, 2009, https://www.psychologytoday.com/us/blog/the-happiness-project/200910/stop-expecting-change-your-habit-in-21-days.

47. David Biello, "Fact or Fiction? Archimedes Coined the Term 'Eureka' in the Bath," *Scientific American,* December 8, 2006, https://www.scientificamerican.com/article/fact-or-fiction-archimede.

48. Benjamin Baird et al., "Inspired by Distraction: Mind Wandering Facilitates Creative Incubation," *Psychological Science* 23, no. 10 (2012): 1117–1122, https://doi.org/10.1177/0956797612446024.

49. Toshikazu Kawagoe, Keiichi Onoda, and Shuhei Yamaguchi, "Different Pre-Scanning Instructions Induce Distinct Psychological and Resting Brain States during Functional Magnetic Resonance Imaging," *European Journal of Neuroscience* 47, no. 1 (2018): 77–82, https://doi.org/10.1111/ejn.13787.

50. Malia F. Mason et al., "Wandering Minds: The Default Network and Stimulus-Independent Thought," *Science* 315, no. 5810 (2007): 393–395, https://doi.org/10.1126/science.1131295.

51. Kalina Christoff et al., "Experience Sampling during fMRI Reveals Default Network and Executive System Contributions to Mind Wandering," *Proceedings of the National Academy of Sciences of the United States of America* 106, no. 21 (2009): 8719–8724, https://doi.org/10.1073/pnas.0900234106; Akina Yamaoka and Shintaro Yukawa, "Does Mind Wandering during the Thought Incubation Period Improve Creativity and Worsen Mood?" *Psychological Reports* 123, no. 5 (2020): 1785–1800, https://doi.org/10.1177/0033294119896039.

52. Timothy D. Wilson et al., "Just Think: The Challenges of the Disengaged Mind," *Science* 345, no. 6192 (2014): 75–77, https://doi.org/10.1126/science.1250830.

53. "Meditation: A Simple, Fast Way to Reduce Stress," Mayo Clinic Patient Care and Health Information, October 17, 2017, https://www.mayoclinic.org/tests-procedures/meditation/in-depth/meditation/art-20045858; Elizabeth A. Hoge et al., "The Effect of Mindfulness Meditation Training on Biological Acute Stress Responses in Generalized Anxiety Disorder," *Psychiatry Research* 262 (2018): 328–332, https://doi.org/10.1016/j.psychres.2017.01.006; Anup Sharma et al., "A Breathing-Based Meditation Intervention for Patients with Major Depressive Disorder Following Inadequate Response to Antidepressants," *Journal of Clinical Psychiatry* 78, no. 1 (2017): e59–e63, https://doi.org/10.4088/JCP.16m10819.

54. Gunes Sevinc et al., "Common and Dissociable Neural Activity after Mindfulness-Based Stress Reduction and Relaxation Response Programs," *Psychosomatic Medicine* 80, no. 5 (2018): 439–451, https://doi.org/10.1097/PSY.0000000000000590; Yi-Yuan Tang et al., "Short-Term Meditation Increases Blood Flow in Anterior Cingulate Cortex and Insula," *Frontiers in Psychology*, February 25, 2015, https://doi.org/10.3389/fpsyg.2015.00212; Kieran C. R. Fox et al., "Functional Neuroanatomy of Meditation: A Review and Meta-Analysis of 78 Functional Neuroimaging Investigations," *Neuroscience and Biobehavioral Reviews* 65 (2016): 208–228, https://doi.org/10.1016/j.neubiorev.2016.03.021.

55. Gregory N. Ruegsegger and Frank W. Booth, "Health Benefits of Exercise," *Cold Spring Harbor Perspectives in Medicine* 8, no. 7 (2018): a029694, https://doi.org/10.1101/cshperspect.a029694; Christiano R. Alves et al., "Influence of Acute High-Intensity Aerobic Interval Exercise Bout on Selective Attention and Short-Term Memory Tasks," *Perceptual and Motor Skills* 118, no. 1 (2014): 63–72, https://doi.org/10.2466/22.06.PMS.118k10w4.

56. Satoshi Hattori, Makoto Naoi, and Hitoo Nishino, "Striatal Dopamine Turnover during Treadmill Running in the Rat: Relation to the Speed of Running," *Brain Research Bulletin* 35, no. 1 (1994): 41–49, https://doi.org/10.1016/0361

-9230(94)90214-3; Christopher M. Olsen, "Natural Rewards, Neuroplasticity, and Non-Drug Addictions," *Neuropharmacology* 61, no. 7 (2011): 1109–1122, https://doi.org/10.1016/j.neuropharm.2011.03.010.

10. Is There Hope?

1. James R. Flynn, "Massive IQ Gains in 14 Nations: What IQ Tests Really Measure," *Psychological Bulletin* 101, no. 2 (1987): 171–191, http://dx.doi.org/10.1037/0033 -2909.101.2.171.

2. James R. Flynn, "Reflections about Intelligence over 40 Years," *Intelligence* 70 (2018): 73–83, https://doi.org/10.1016/j.intell.2018.06.007.

3. James Flynn, "Why Our IQ Levels Are Higher than Our Grandparents'," TED, March 2013, https://www.ted.com/talks/james_flynn_why_our_iq_levels_are _higher_than_our_grandparents.

4. Lisa Trahan et al., "The Flynn Effect: A Meta-Analysis," *Psychological Bulletin* 140, no. 5 (2014): 1332–1360, https://doi.org/10.1037/a0037173; Bernt Bratsberg and Ole Rogeberg, "Flynn Effect and Its Reversal Are Both Environmentally Caused," *Proceedings of the National Academy of Science of the United States of America* 115, no. 26 (2018): 6674–6678, https://doi.org/10.1073/pnas.1718793115.

5. Monica Anderson and Jingjing Jiang, "Teens' Social Media Habits and Experiences," Pew Research Center, November 28, 2018, http://www.pewinternet.org /2018/11/28/teens-social-media-habits-and-experiences.

6. Jay N. Giedd, "The Digital Revolution and Adolescent Brain Evolution," *Journal of Adolescent Health* 51, no. 2 (2012): 101–105, https://doi.org/10.1016/j.jadohealth .2012.06.002.

7. "Internet-Broadband Fact Sheet," Pew Research Center, Internet and Technology, February 5, 2018, http://www.pewinternet.org/fact-sheet/internet-broadband.

8. "Number of Mobile Phone Users Worldwide from 2015 to 2020 (in Billions)," *Statista*, August 2, 2018, https://www.statista.com/statistics/274774/forecast-of -mobile-phone-users-worldwide.

9. Kathryn Zickuhr and Aaron Smith, "Digital Differences," Pew Research Center, April 13, 2012, http://www.pewinternet.org/2012/04/13/digital-differences.

10. Emily Olsen, "Digital Health Apps Balloon to More than 350,000 Available on the Market, According to IQVIA Report," *MobiHealthNews*, August 4, 2021, https://www.mobihealthnews.com/news/digital-health-apps-balloon-more -350000-available-market-according-iqvia-report; Pooja Chandrashekar, "Do Mental Health Mobile Apps Work: Evidence and Recommendations for Designing High-Efficiency Mental Health Apps," *mHealth* 4 (2018): 6, https://doi .org/10.21037/mhealth.2018.03.02.

11. Sarah Perez, "Apple Unveils a New Set of 'Digital Wellness' Features for Better Managing Screen Time," *TechCrunch,* June 4, 2018, https://techcrunch.com/2018 /06/04/apple-unveils-a-new-set-of-digital-wellness-features-for-better -managing-screen-time.

12. Tracey J. Shors, "The Adult Brain Makes New Neurons, and Effortful Learning Keeps Them Alive," *Current Directions in Psychological Science* 23, no. 5 (2014): 311–318, https://doi.org/10.1177/0963721414540167.

ACKNOWLEDGMENTS

I want to start with thanks to the many scholars who work tirelessly to answer the difficult questions addressed in this book. Having come from the world of scientific research, I know the challenges, applaud the effort, and am grateful for the opportunity to share the fruits of this hard labor with an audience beyond the realm of academic journal articles, lecture halls, and conference rooms.

I also wish to thank my clients and colleagues in media and market research. When I entered the industry in 2006, I thought of marketers as authorities on consumer desires who used rational arguments to persuade people to buy things. But my clients, many of whom worked for Fortune 500 brands, helped me think differently by embracing new neuroscientific insights and applying them to their business endeavors. These pioneers in consumer neuroscience were my partners in learning about consumer behavior. Over time, I saw a massive shift in my own thinking and that of the industry. The shift was from a form of advertising that emphasized authority to one that emphasized authenticity; from rational to emotional; and from attempts to persuade to attempts to engage. This is what we have learned from the neuroscience of consumers. Consumers want to be free to choose what is best for them. They don't want to be manipulated. Arriving at this point required risk-taking and a willingness to test the once-radical notion that neuroscience could deepen our understanding of consumer behavior. I offer my sincere thanks to the market and media researchers who had the courage to go deeper.

I also want to thank the many journalists who have helped bring neuroscience research to the attention of the public. Their ability to distill complex neurobiological findings into an accessible and digestible form is a constant inspiration. Many journalists separately interviewed the researchers I cite, providing a bounty of quotations and information.

There is one journalist in particular for whom I'm especially grateful: Joe Mandese of the publisher MediaPost. In 2011, after many rounds of coffee and conversation, Joe asked me to be guest editor of a special edition of *Media Magazine,* one of MediaPost's publications. For that project, Joe and I wrote many stories together about the proliferation of screens, the rise of media multitasking, and the potential of neuroscience to increase our understanding of consumer behaviors. We covered music, nostalgia, and advertising through the lens of neuroscience and the brain. It was an amazing experience working with a professional editor. We produced a unique work integrating neuroscientific concepts with those from a more traditional marketing and media perspective. Joe encouraged me to write this book. For that and for his wisdom and his willingness to listen, I am grateful.

Then there are my incredible friends, who helped me navigate the cryptic path to publishing. Stephen and Liz Kendrick and Jim Whitters were supporters of this book from conception to final publication. They were an amazing complement to the very capable editors at Harvard University Press, the thoughtful professional reviewers, and my other informal reviewers, whose comments on early drafts were also invaluable. In every case these informal reviewers were friends who are also parents—parents who share my concerns about the challenges of raising children in the digital age. Making complex science accessible to a general audience is difficult, as is the task of inspiring parents to think differently about digital literacy and the consequences, for themselves and their loved ones, of misusing mobile media, communications, and information technology. It would have been impossible to write this book without feedback from contemplative parents who are worried about their children's mental health.

And to my wife, Jeannine, who stood by my side when I left academic medicine to pursue entrepreneurship, who worked for and with me in the nascent field of consumer neuroscience and partners with me in raising our three beautiful children. Jeannine and I speak often about our family media plan; we struggle as much as any parents to raise children in a world saturated with mobile technologies that were unimaginable when we were kids. Her keen powers of observation helped me develop some of the ideas in this book. For that, I am grateful to her.

Finally, I am grateful to the Aspen Institute and all the members of the 2014 "Bones & Elephants" class of Henry Crown Fellows. The Aspen Global Leadership Network is full of inspiring people doing great work that has a positive impact on the planet. To a person, the members of my class have supported me and pushed me to do my best work, sometimes softly and, when I needed it, more forcefully. This book is a direct result of the Aspen program and dialogue with friends and fellows there. Thank you.

INDEX